Regulation and Dysfunction of Apoptosis

Edited by Yusuf Tutar

Published in London, United Kingdom

IntechOpen

Supporting open minds since 2005

Regulation and Dysfunction of Apoptosis
http://dx.doi.org/10.5772/intechopen.92915
Edited by Yusuf Tutar

Contributors
Rafiq A. A Rather, Sonia Thapa, Madhulika Bhagat, Shashank K. Singh, Namasivayam Nalini, Bichandarkoil Jayaram Pratima, Hilal Kalkan, Raju Mondal, Sreya Antony, Sovan Roy, Sanjib Kumar Chattopadhyay, Chang-Hui Shen, Goldie Libby Sherr, Dae Joon Kim, Bilal Bin Hafeez, Eunmi Park, Kyung-Soo Chun, Yong-Yeon Cho, Reyhaneh Farghadani, Rakesh Naidu, Yusuf Tutar, Mehmet Evren Okur, Panoraia I. Siafaka, Merve Tutar

First published in London, United Kingdom, 2022 by IntechOpen
IntechOpen is the global imprint of INTECHOPEN LIMITED, registered in England and Wales, registration number: 11086078, 5 Princes Gate Court, London, SW7 2QJ, United Kingdom
Printed in Croatia

British Library Cataloguing-in-Publication Data
A catalogue record for this book is available from the British Library

Additional hard and PDF copies can be obtained from orders@intechopen.com

Regulation and Dysfunction of Apoptosis
Edited by Yusuf Tutar
p. cm.
Print ISBN 978-1-83969-193-5
Online ISBN 978-1-83969-194-2
eBook (PDF) ISBN 978-1-83969-195-9

For EU product safety concerns:
IN TECH d.o.o., Prolaz Marije Krucifikse Kozulić 3, 51000 Rijeka, Croatia, info@intechopen.com or visit our website at intechopen.com.

Meet the editor

Prof. Dr. Yusuf Tutar is currently a principal investigator at the Hamidiye Faculty of Pharmacy, Department of Basic Pharmaceutical Sciences, Division of Biochemistry, University of Health Sciences, Turkey,. He is also a faculty member in the Molecular Oncology Program. He obtained his MSc and Ph.D. at Oregon State University and Texas Tech University, respectively. He pursued his postdoctoral studies at Rutgers University Medical School, New Jersey, and the National Institutes of Health (NIH), USA. His research focuses on biochemistry, biophysics, genetics, molecular biology, and molecular medicine with specialization in the fields of drug design, protein structure-function, protein folding, prions, microRNA, pseudogenes, molecular cancer, epigenetics, metabolites, proteomics, genomics, protein expression, and characterization by spectroscopic and calorimetric methods.

Contents

Preface

The pathogenesis of many diseases is most closely related to inappropriate apoptosis. For example, cancer is one of the situations where too little apoptosis leads to the development of malignant cells that highly proliferate. Defects at any points along apoptotic pathways may lead to malignant transformation of the affected cells, tumor metastasis, and resistance to anti-cancer drugs. Strategies targeting the master regulators in carcinoma cells have been a major focus of interest in cancer studies. Therefore, despite being the cause of the problem, apoptosis can be targeted in cancer therapy.

This book begins with Chapter 1, "The Role of Apoptosis as a Double-Edge Sword in Cancer," which provides a comprehensive review of apoptotic cell death and examines how deficiencies in apoptotic master regulators, caspases, and Bcl-2 family proteins influence carcinogenesis and can be targeted in cancer treatment. Chapter 2, "The Program Cell Death (Apoptosis) and The Therapy of Cancer," deals with apoptosis's vital roles in maintaining organ homeostasis. Programmed cell death occurs when the DNA damage is irremediable and has two important pathways: the intrinsic death pathway, also known as the mitochondrial pathway, and the extrinsic programmed cell death pathway. Any defects in the regulation of these crucial pathways have been associated with many disorders, most importantly cancer. Therefore, understanding the molecular basis of apoptosis is essential for the treatment of incurable cancer. Hence, it is important to understand the maintenance and counteraction of apoptosis and develop and improve successful new pharmacological applications of cell death mechanisms for future therapies. This chapter discusses the mechanism of apoptosis and emerging principles of drug resistance in cancer. Chapter 3, "Recent Advancements in Apoptosis-Based Therapeutic Approaches for Cancer Targeting," discusses the development of novel drug targets and drug delivery systems for inhibiting or inducing apoptosis. Apoptosis plays a major role in cellular homeostasis, normal development, and clearance of cells. Non-programmed cell death can take place by various external factors, such as infection, toxins, and physical injury. The chapter provides recent insights into therapeutic approaches to cancer. Chapter 4, "Insights into the Role of Defective Apoptosis in Cancer Pathogenesis and Therapy," discusses defective apoptosis as an indispensable causative factor in the development of cancer that allows cancer cells to survive longer and favors the accumulation of oncogenic mutations. Further, upregulation of anti-apoptotic proteins and loss of pro-apoptotic proteins strongly favors apoptosis evasion. The ability of cancer cells to evade apoptosis is critical for the progression and clonal expansion of malignantly transformed cells. Defective apoptosis imparts proliferative advantage to cancer cells or cells with the potential to become cancerous. The mechanisms employed by cancer cells to evade apoptosis can be used in the strategic design of therapeutic regimens aimed at exploiting apoptotic signaling networks to ensure tumor-specific cell death. This chapter presents knowledge about the molecular mechanisms of defective apoptosis that could be translated into the development of novel therapeutic agents and therapeutic modalities for cancer treatment. Chapter 5, "Efferocytosis: An Interface between Apoptosis and Pathophysiology," deepens our understanding of cellular death mechanism. Cell death occurring under physiological conditions is primarily caused by apoptosis, which is a non-inflammatory or silent process, whereas necroptosis or pyroptosis is triggered by pathogen invasion, which stimulates the immune system and induces inflammation. In physiology, clearing dead cells and associated cellular debris is necessary since billions of cells die during mammalian embryogenesis and every day in adult organisms. For degradation, dead cells produced by apoptosis are quickly engulfed by macrophages. This chapter presents a description of the phagocytosis of dead and dying cells, as a process known as efferocytosis. Chapter 6, "Programmed Cell Death (PCD) in Plant: Molecular Mechanism, Regulation and Cellular Dysfunction in Response to Development and Stress," highlights the current understanding of plant programmed cell death in terms of different pathways, cellular responses, regulation, and signaling mechanisms. The study helps to understand the molecular and structural instincts of programmed cell death in different stages of plant growth and development, response to biotic/abiotic stimuli, and cellular dysfunction. Chapter 7, "Regulation of Apoptosis during Environmental Skin Tumor Initiation," discusses skin cancer. Skin cancer has three distinct stages: initiation, promotion, and progression. During the initiation, the fate of DNA-damaged skin cells is determined by the homeostatic regulation of pro-apoptotic and anti-apoptotic signaling pathways. The imbalance or disruption of either signaling will lead to the survival of initiated cells, resulting in the development of skin cancer. This chapter elaborates on the signaling pathways that regulate apoptosis and the impact of their dysfunction during skin tumor initiation. Finally, Chapter 8, "The Interplay of Key Phospholipid Biosynthetic Enzymes and the Yeast V-ATPase Pump and Their Role in Programmed Cell Death," covers a model organism apoptotic mechanism. Exposure of the yeast *Saccharomyces cerevisiae* to environmental stress can influence cell growth, physiology, and differentiation, and thus result in cells' adaptive response. During the course of adaptive response, the yeast vacuoles play an important role in protecting cells from stress. The vacuole is a dynamic organelle and is similar to lysosomes in mammalian cells. The defect of the lysosome's function may cause various genetic and neurodegenerative diseases. The multi-subunit V-ATPase is the main regulator for vacuolar function and its activity plays a significant role in maintaining pH homeostasis. V-ATPase is an ATP-driven proton pump required for vacuolar acidification. It has also been demonstrated that phospholipid biosynthetic genes might influence vacuolar morphology and function. However, the mechanistic link between phospholipid biosynthetic genes and vacuolar function has not been established. Recent studies have demonstrated the regulatory role of Pah1p, a phospholipid biosynthetic gene, in V-ATPase disassembly and activity. Therefore, this chapter employs *S. cerevisiae* as a model to show how Pah1p affects V-ATPase disassembly and activity and how Pah1p negatively affects vacuolar function.

Dr. Yusuf Tutar
Professor,
Department of Basic Pharmaceutical Sciences,
Division of Biochemistry,
Hamidiye Faculty of Pharmacy,
University of Health Sciences-Turkey,
Istanbul, Turkey

Section 1

General Mechanism of Apoptosis

Chapter 1

The Role of Apoptosis as a Double-Edge Sword in Cancer

Reyhaneh Farghadani and Rakesh Naidu

Abstract

The pathogenesis of many diseases is most closely related to inappropriate apoptosis (either too little or too much) and cancer is one of the situations where too little apoptosis happens, leading to malignant cells that highly proliferate. Defects at any points along apoptotic pathways may lead to malignant transformation of the affected cells, tumor metastasis, and resistance to anti-cancer drugs. Several major molecular mechanisms are involved in the evasion of apoptosis in cancer initiation and progression. Bcl-2 family of proteins and caspases are the central players in the apoptotic mechanism and regulate cell death. Their imperfections cause to the deficient apoptotic signaling and thereby the inadequate apoptosis in cancer cells and eventually carcinogenesis. Strategies targeting these master regulators in carcinoma cells has been a major focus of interest in cancer studies. Therefore, despite being the cause of problem, apoptosis can be targeted in cancer therapy. This chapter provides a comprehensive review of apoptotic cell death and how deficiencies in apoptotic master regulators, caspases and Bcl-2 family proteins, influence carcinogenesis and can be targeted in cancer treatment.

Keywords: apoptosis, cancer, Bcl-2, caspase, regulation, dysfunction, intrinsic pathway, extrinsic pathway, carcinogenesis

1. Introduction

Cancer as a complicated and heterogeneous disorder is the major threat to human beings and is still the significant leading cause of mortality around the world. According to the world health organization report, cancer is the second leading cause of death around the world with 9.6 million deaths in 2018. That is nearly 1 in 6 of all global deaths [1, 2]. The incidence of cancer is expected to rise approximately 70% within the next two decades around the world, from 14 million new cases in 2012 to 25 million new cases a year [3–5]. Cancer development comprises of a multiple steps happening progressively and beginning with initial mutations that promote tumorigenesis and, eventually, metastasis. The genetic alterations ultimately cause to a disturbance in the balance between cell proliferation and programmed cell death or apoptosis [6].

Apoptosis is a process of the cell's natural mechanism for death which occurred in multicellular organisms to maintain tissue homeostasis and act as a defensive strategy to remove infected, damaged or mutated cells. Apoptosis can be triggered through two major pathways, either mitochondrial- or death receptor-mediated pathways resulting from the intracellular (e.g. stress, DNA damage) and extracellular signals (death-inducing signals by other cells), respectively. This machinery

 IntechOpen

mainly depends on caspases cascades for executing cell death that eventually cause proteolytic cleavage of thousands of target proteins within the cells that are essential for normal cellular function such as cytoskeletal and nuclear proteins. Consequently, the apoptotic cells undergo a series of morphological and biochemical alterations leading to recognition by macrophages and cell phagocytosis. Moreover, B-cell lymphoma-2 (Bcl-2) family of proteins has long been identified for their significant involvement in regulating the cellular program of apoptosis through mitochondrial outer membrane permeabilization, as the critical decision-point at which cells commit to death, representing their vital role in protecting against cancer [7–9].

Deficiencies at any point along apoptotic pathways and dysfunction of the controlling mechanisms may result in impaired apoptosis that cause to carcinogenesis, allowing cancer cells to survive over intended lifespans and eventually uncontrolled cell proliferation, tumor development and progression. Tumor cells evade apoptosis through a variety of mechanisms. Understanding these molecular mechanisms not only provide insight into the cancer pathogenesis, but also provide clues on cancer treatment [7, 10]. Besides, genomic instability, nutrient deficiency, cellular hypoxia and oncogenic stress may cause to continuous stress within cancer cells which make them more sensitive to apoptotic stimulation. Hence, the ability to target the molecular components of this machinery and restore an apoptotic pathway has long been considered as an intriguing approach in cancer drug discovery. Consequently, being as a double-edged sword, apoptosis plays a critical role in both tumorigenesis and cancer therapy [6, 11, 12]. Therefore, as evasion of apoptosis is well known as the hallmark of all types of cancers, this chapter will be mainly emphasizing the role of apoptosis in cancer, from pathogenesis and cancer development to cancer therapy and treatment with primarily focus on two key mediators of apoptosis, caspases and Bcl-2 family of proteins, which have been receiving great attention in targeted cancer therapies.

2. The role of apoptosis in pathogenesis and treatment of cancer

2.1 Overview of apoptosis

The term "apoptosis" is derived from Greek, meaning "dropping off" or "falling off" as leaves from a tree, was first used in 1972 to describe a morphologically distinct form of cell death. Apoptosis also known as programmed cell death is a highly regulated energy-dependent process that occurs normally during development and aging. It plays an important role as a homeostatic mechanism to maintain cell populations in the tissue of multicellular organisms. In addition to its importance in biological process, defects in apoptosis mechanism has been implicated in the pathophysiology of diseases including cancer [13, 14]. There are many factors, mostly proteins, involved in the activation and regulation of apoptotic mechanism. This highly complicated and regulated process involves an energy-dependent cascade of molecular events and includes the mitochondria-dependent (intrinsic) and death receptor-dependent (extrinsic) pathway. Caspases play a vital role in initiation and execution of both intrinsic and extrinsic pathways which is mediated through the cleavage of hundreds of proteins essential for normal cellular function [15].

2.1.1 Caspases: key apoptotic proteins

Caspases are a family of cysteine protease enzymes that are able to selectively cleave proteins at aspartic acid residues using the sulfur atom of cysteine in their catalytic site, hence, named as cysteine-aspartic proteases or caspases. They play an essential role in maintaining homeostasis through regulating cell death and inflammation. Caspases have been generally categorized by their known functions in apoptosis (caspase-2, -3, -6, -7, - 8, -9 and -10) and in inflammation (caspase-1, -4, -5 and -12) in human. Caspases involved in apoptotic cell death have been then subgrouped based on their position and mechanism of action in apoptotic signaling cascades and are either initiator caspases (caspase-8, -9 and -10) or executioner caspases (caspase-3, -6, and -7) in apoptotic pathway. Therefore, caspases as a conserved family of cysteine proteases, which are essential in initiation and execution of intrinsic and extrinsic pathways, are the main emphasis of apoptosis studies [16–18].

Caspases are initially synthesized as an inactive monomeric proenzyme, named zymogens or procaspases, containing a large subunit, small subunit, and prodomain that is only activated through proteolytic cleavage and dimerization following an appropriate stimulus. Therefore, this post-translational level of control provides rapid and tight regulation of the caspase enzyme activities [19, 20]. Initiator caspases have prodomains containing one of the two specific protein–protein interaction domain including caspase recruitment domain (CARD) and death effector domain (DED) that promote caspase dimerization through binding to adapter proteins. Two examples of activating multiprotein complexes include death-inducing signaling complex (DISC) and the apoptosome, which are formed during extrinsic and intrinsic pathway of apoptosis, respectively [19].

Once properly assembled into dimers, pro-caspases undergo cleavage by autocatalysis resulting in the removal of pro-domain and cleavage at the linker region between the large and small subunit resulting in the heterotetramer formation and provides the active-site loops to get a proper conformation for enzymatic activity [17, 19].

Although, initiator caspases are capable of autocatalytic cleavage and activation, effector caspases are cleaved by initiator caspases resulting in the formation of active heterotetramer. Each active caspase is a tetramer consists of two identical big subunits and two identical small subunits. Accordingly, activation of apoptotic caspases leads to the inactivation or activation of substrates, and therefore initiation of a protease cascade events in the apoptotic signaling pathway resulting in rapid cell death. Activated caspases trigger cytoplasmic endonuclease, cleave many vital cellular proteins and break up the nuclear scaffold and cytoskeleton as well as activate DNase, which further degrade nuclear DNA into 180 to 200 base pair. Collectively, caspase activity results in various morphological and biochemical changes in apoptotic cells [19, 21, 22].

2.1.2 Morphological changes in apoptosis

Apoptotic cells are differentiated from healthy and necrotic cells based on certain cellular morphological changes. Characteristic features of apoptosis in the nucleus are chromatin condensation and nuclear fragmentation which are accompanied by cell shrinkage, membrane blebbing and formation of apoptotic bodies in the final stage of apoptosis which are rapidly engulfed by phagocytosis that avoids an inflammatory response in surrounding tissues [23–25]. The shrinkage of the cell is one of the most common morphological changes in apoptotic cell death resulted from the extreme alteration in intracellular water. Intracellular water plays a critical role in apoptotic and necrotic cell death. Although necrotic cells absorb the water resulting in enlarging the size and finally burst, apoptotic cells lose water leads to reduction in cellular volume. Therefore, the apoptotic cells become smaller in size, the cytoplasm is dense and the organelles are more tightly packed. Consequently, due to the occurrence of cell shrinkage, the cell will lose its contact with neighboring cells, or the extracellular matrix and acquire more rounded morphology. Although the plasma membrane is intact during the entire process, at later stage of apoptosis, loss of membrane integrity and formation of the blebs at the cell surface due to the separation of the plasma membrane from cytoskeleton occur in apoptotic cells [26–28].

2.1.3 Biochemical changes in apoptosis

Apart from structural alterations, several biochemical changes also play key events in apoptosis. Apoptotic cells generally display major types of biochemical modifications such as caspase activation, protein and DNA cleavage, and plasma membrane alterations, which lead to phagocytic recognition [13]. Disruption of plasma membrane asymmetry is a common feature of apoptotic cells, independent of the form of apoptotic stimulus. The maintenance of lipid asymmetry of plasma membrane is regulated through transporters named flippases and floppases. In addition, the activated scramblase enzymes have an important role in the loss of lipid asymmetry and enhanced phosphatidylserine (PS) exposure to the outer leaflet of plasma membrane [13, 29].

Therefore, in a healthy cell, PS is limited to the inner layer of the plasma membrane. However, during apoptosis, effector caspases cleave and activate scramblase, as well as cleave and inactivate flippase, responsible for transmitting PS from the outer to the inner leaflet that lead to externalization of PS. Therefore, phosphatidylserine, which is normally localized in the inner membrane layer of cells is flipped out and externalized on the outer layer of the plasma membrane. This PS externalization not only is the indicator of loss of membrane asymmetry during apoptosis, but also allows early recognition by phagocytes and prevents the release of proinflammatory cellular components [29–31].

2.1.4 Pathway of apoptosis

As mentioned earlier, the mechanism of apoptosis involves an energy-dependent cascade of molecular events. Apoptotic cell death machinery includes the mitochondria-dependent (intrinsic) pathway and death receptor-dependent (extrinsic) pathway. The intrinsic pathway arises from intracellular signals like cellular stress and DNA damage and relies on the release of proteins from the intermembrane space of mitochondria. However, the extrinsic pathway is activated through the binding of extracellular ligands to death receptors at the cell surface that trigger the multiprotein complex formation known as death-inducing signaling complex (DISC). These two mitochondria- and death receptor-mediated pathways are interconnected and the molecules in one pathway can affect another pathway [32, 33].

2.1.5 The intrinsic mitochondrial pathway

As its name implies, the intrinsic pathway is activated in response to internal stimuli such as hypoxia, severe DNA damage and oxidative stress and mitochondria play a critical role throughout this apoptosis signaling pathway [34, 35]. The intrinsic pathway is mainly controlled by the members of Bcl-2 family proteins, which regulate the permeabilization of mitochondrial outer membrane (MOM) and are structurally and functionally classified into three groups. BH3-only proteins, like Bim and Bik, that sense cellular stress and directly or indirectly activate the executioner proteins, like Bax, Bak, Bid, that are able to oligomerize in and permeabilize the MOM. The oligomerization of these pro-apoptotic proteins leads to component release form the intermembrane space to the cytoplasm and activation of effector caspases of apoptosis. The first two groups are known as the pro-apoptotic proteins of Bcl-2 family. The third group is the anti-apoptotic proteins, like Bcl-2 and Bcl-xL that hinder the overall process by inhibiting pro-apoptotic proteins. However, not the absolute quantity but rather the relative levels and balance between the pro- and anti-apoptotic proteins regulates whether the apoptosis event would be initiated. Although the excess of pro-apoptotic proteins makes the cells sensitive to apoptosis, excess of anti-apoptotic proteins makes the cells resistant and prevents the occurrence of apoptosis [36–38]. However, in the presence of apoptotic stimuli, the death signal is sensed initially by the BH3-only protein, which then interacts with the downstream mediators of apoptosis such as Bax. As the intrinsic mitochondrial pathway is initiated, Bax is translocated from cytosol to the outer mitochondrial membrane. The assembly of Bax proteins in mitochondrial outer membrane results in protein-lined channels or pore formation and intensely increase its permeability that cause a dramatic loss of electrical potential in mitochondria and cytochrome c release to cytoplasm. Subsequently, released cytochrome c binds to APAF-1 to facilitate the formation of the apoptosome, a wheel shaped heptametrical complex, which can then recruit and activate pro-caspase-9. Consequently, caspase-9 activates effector caspases (caspase-3/-7) that eventually lead to apoptosis (**Figure 1**) [39–41].

2.1.6 The extrinsic death receptor pathway

The extrinsic pathway is activated through the interactions between the transmembrane death receptors of the tumor necrosis factor (TNF) superfamily and their related ligands. The TNF receptor family has common cysteine-rich extracellular domains and cytoplasmic death domains that involve in transmitting the death signal from the cell surface to the intracellular signaling pathways. Ligation of death receptors with death ligands causes conformational change in death domain and consequently recruits apoptosis-related adaptor proteins that associate with procaspase-8/-10. At this point, a death-inducing signaling complex (DISC) consisting of the death receptor, an adaptor molecule, and pro-caspase-8/−10 is formed, resulting in the auto-catalytic activation of procaspases (**Figure 1**). The activated form of the caspase-8/-10 enzyme, as an initiator caspase, subsequently cleaves and activates other downstream or executioner caspases [42, 43]. Finally, both apoptotic pathways result in the activation of effector caspases (caspase-3/-7) causing the cleavage of key cellular macromolecules which are required for normal cellular function. They cleave the structural proteins in the cytoskeleton and nuclear proteins such as DNA repair enzymes and activate degradative enzymes such as DNases, which together contribute to the typical morphological changes and promote cell death [44, 45].

2.2 Dysregulation of apoptosis in carcinogenesis

The pathogenesis of many diseases is most closely related to inappropriate apoptosis (either too little or too much) and cancer is one of the situations where too little apoptosis happens, leading to malignant cells that highly proliferate. Defects at any points along apoptotic pathways may lead to malignant transformation of the affected cells, tumor metastasis, and resistance to anti-cancer drugs [12, 46]. Defects in Bcl-2 family of proteins and caspases are well-known chief factors to be involved in the evasion of apoptosis by tumor cells.

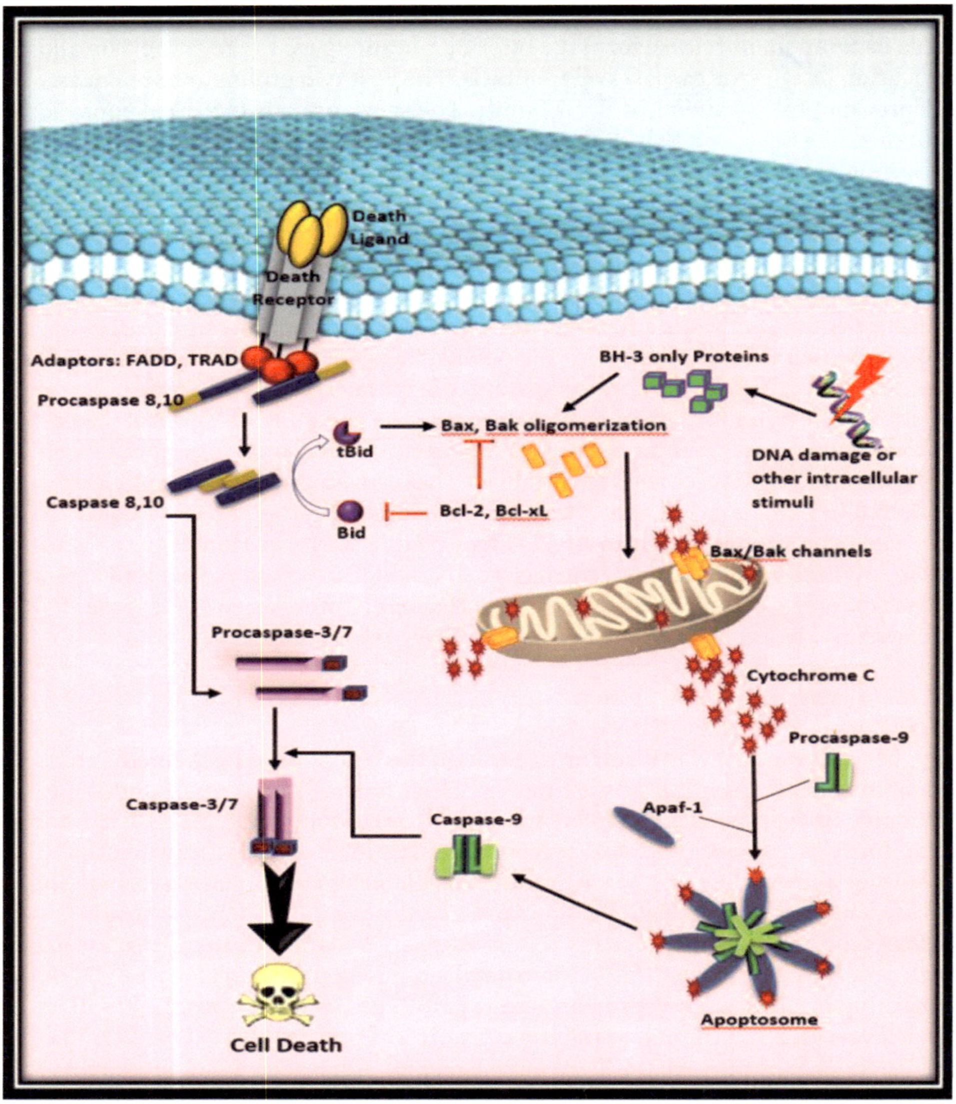

Figure 1.
Apoptosis signaling pathways. Abbreviations: TRADD, TNF receptor-associated death domain protein; FADD, Fas-associated death domain protein; Bid, BH3 interacting-domain death agonist; Bak, Bcl-2 homologous antagonist/killer; tBid, truncated BID; Bax, Bcl-2 associated X protein; APAF-1, apoptotic protease activating factor-1; Bcl-2, B-cell lymphoma 2, Bcl-xL; B-cell lymphoma-extra large.

2.2.1 Dysfunction of the Bcl-2 family of proteins in apoptosis

The Bcl-2 family of proteins consist of pro-apoptotic and anti-apoptotic proteins that act as a master regulator of initiation of apoptosis through intrinsic pathway and function chiefly at the mitochondrial level. The first protein of this family, B-cell lymphoma 2 (Bcl-2), was recognized almost 30 years ago. Currently 25 members of the Bcl-2 family have been determined and based on the presence of conserved Bcl-2 homology (BH) domains and their role in mitochondrial-mediated apoptosis, they are categorized into the following three subfamilies [47, 48]. Anti-apoptotic subgroup consisting of Bcl-2, Bcl-w, Bcl-xL, A1/Bfl-1, Mcl-1and Bcl-B/Bcl2L10 proteins contain four BH domains designated as 1–4 and inhibit the apoptosis occurrence so named pro-survival proteins. However, second group, known as apoptosis effectors, belongs to pro-apoptotic members of this family containing BH 1–3 and missing the BH4 domain. Some example of this group are Bak, Bax, and Bok/Mtd. The last group that can be considered as subdivision of pro-apoptotic proteins including Bik, Bid, Bim, Bmf, Puma, Bad, Hrk and Noxa are named "BH3-only"' proteins as they contain only the BH3 domain. The members of this group function as initiators of apoptosis and the major mediators of the interaction with anti-apoptotic proteins [47, 49, 50]. Structural studies have determined that BH1, BH2 and BH3 areas together form a hydrophobic pocket that can be filled by the amphipathic a-helical BH3 domain of pro-apoptotic Bcl-2 proteins. Consequently, Bcl-2 family interactions regulate mitochondrial outer membrane (MOM) integrity and function and eventually onset of mitochondrial-mediate apoptosis [37, 51].

The balance disturbance of anti-apoptotic and pro-apoptotic proteins cause to dysregulated apoptosis in the affected cells. Altered expression of these proteins frequently occurs in cancers. Overexpression of anti-apoptotic proteins such as Bcl-2 or Bcl-xL occurs in a huge number of human cancers [52–55]. In one study, targeted proteomic analysis have revealed the contribution of Bcl-2 overexpression to cell survival of laryngeal carcinoma (LC) though its interaction with Hsp90β and formation of Bcl-2 Hsp90β complex involving in the anti-apoptotic progression of LC [56]. In cervical cancer SiHa cells, overexpressing Bcl-2 gene, the suppression of down-regulation of HPV16 E7 and c-myc gene expression may inhibit the induction of apoptosis [57]. Besides, high levels of Bcl-2 have been reported in hematological malignancies. Various mechanisms such as gene amplification, chromosomal translocations and dysregulation of miRNAs involved in Bcl-2 RNA degradation may cause to Bcl-2 upregulation [58–60]. Furthermore, there have been a number of studies reporting the involvement of Bcl-xL anti-apoptotic protein in tumorigenesis. The increased level of Bcl-xL gene expression determined in human cancers such as colorectal cancer, breast cancer, gastric adenomas and carcinomas, hepatocellular carcinoma and prostate cancer promotes cancer cell survival [61–65]. In addition, several attempts have revealed the association of enhanced levels of Bcl-xL and MCL1 with the malfunction of miRNAs that usually diminish their expression such as miR-29, miR-125b, miR-193 [66–68]. Furthermore, overexpression of anti-apoptotic Bcl-2 and its close relatives have been recognized as chief components of chemoresistance [69–72].

Deficiency in pro-apoptotic members of the Bcl-2 family has also been extensively studied in tumorigenesis and cancers. Pro-apoptotic gene Bim is frequently silenced in human Burkitt's lymphoma [73, 74]. Homozygous deletion and the loss of mRNA and protein expression have also been determined in mantle cell lymphoma and renal cell carcinoma. Hence, blocking Bim expression caused by gene deletion or epigenetic silencing is mainly contributed to pathogenesis of these tumors [75, 76]. Furthermore, a number of researchers have reported that down-regulation and mutation of Bax plays a significant role in tumor resistance to apoptosis. Reduced Bax expression was reported to be correlated with acquiring resistance to 5-FU in colorectal cancer cell line and cisplatin in ovarian carcinoma [77, 78]. Sensitivity of non-small cell lung cancer to Zoledronic was also found to be Bax dependent [79]. Suppressed Bax activity is one of the major reasons of TRAIL resistance in melanoma [80–82]. Besides, inactivated mutation in gene Bax such as frameshift mutations, loss of function mutations and point mutations has been reported in colon cancers, certain hematopoietic malignancies and acquired resistance to antineoplastic drugs [83–85]. Additionally, cells lacking both Bax and Bak have confirmed to be completely resistant to truncated Bid (t-Bid)-induced cytochrome c release and apoptosis [86]. Therefore, all these abnormalities regarding

Bcl-2 family protein members affect the ratio and equilibrium of pro-apoptotic to anti-apoptotic proteins which result in apoptosis dysfunction and resistance to cell death.

2.2.2 Dysfunction of caspases in apoptosis

Caspases are a family of cysteine proteases that play crucial role in initiation and execution of apoptosis signaling pathway. During tumorigenesis, altered caspase activity or deficiency in their functions may lead to impairing apoptosis induction resulting in intense misbalance in the growth dynamics that eventually cause to decreased apoptosis, irregular growth of cancer cells and carcinogenesis [17, 87]. Human cancer cells dysregulate caspase activity through a different mechanism such as inactivated mutation, down-regulation and epigenetic alteration blocking their apoptotic activity [88–90].

Caspase-3/-7 is a critical executioner molecule in apoptotic mechanism through cleaving a variety of key cellular proteins. Many studies have demonstrated the close association of altered caspase-3 expression and various cancers such as cervical adenocarcinoma, colon cancer, glioma and breast cancer [91–97]. However, the role of caspase-3 in breast cancer patients has been an area of controversy. Meta-analysis study of 3091 cases have revealed that enhanced expression of caspase 3 is related to poor overall survival in patients [98].

As mentioned earlier, the activation of executioner caspases involves their proteolytic cleavage through mature and functioning initiator caspases. Therefore, deficiency in initiator caspases activity has been determined in cancer development and progression [99, 100]. Caspase-9 plays a critical role in the initiation phase of the intrinsic apoptosis pathway. Decreased levels of caspase-9 was reported in patients with stage II colorectal cancer associated with poor clinical outcome [90, 101]. Inhibition of caspase 9 activity has been reported to be involved in acquired cisplatin resistance in head and neck squamous cell carcinoma cells [102]. Several functional polymorphism of caspase-9 has also been determined which may influence its expression or activity and therefore alter susceptibility to cancer [103–106].

Since extrinsic signaling of apoptosis mechanism after external stimulation of the death receptors is mediated through initiators caspase-8 and caspase-10, their deregulated expression or function can block death receptor signaling pathway contributing to cancer development. Expression of caspase-10 was found to be reduced in rectal cancer [107]. The cDNA array analysis has also detected the reduced co-expression of initiator caspases of extrinsic pathway, caspase 8 and 10, that might contribute to the pathogenesis of choriocarcinoma [108]. In previous investigations, expression analysis of caspase-8 has shown its down regulation in breast cancer cell lines and tumor tissues of breast cancer and revealed significant association between altered caspase-8 expression and status of HR in breast cancer patients [109]. Some studies also revealed that loss of caspase-8 expression not only cause to reduced apoptosis, but also involved in enhanced cell migration, tumor microenvironment and amplification of MYCN oncogene which highlight its contribution in carcinogenesis. The lack of caspase-8 expression happens very commonly in neuroendocrine cancers such as glioblastoma, medulloblastoma, neuroblastoma [110–112]. Furthermore, the correlation between caspase-8 with cancer prognosis, cancer stage and therapy resistance has been reported [109, 110]. Loss of initiator caspase-8 protein expression has been shown to be related with undesirable survival outcome in medulloblastoma and gynecological tumors such as ovarian and breast cancers and stage of head and neck squamous cell carcinoma (HNSCC) [113, 114].

2.3 Targeting cellular apoptosis machinery in cancer treatment

Since inhibition of apoptosis lies at the heart of all abnormal malignant growth, metastasis and conferring therapeutic failure, targeting the apoptosis mechanism players is of vital importance in cancer therapy. In this regard, Bcl-2 family of proteins as gate-keepers of intrinsic apoptotic pathway mediating the pro- and anti-apoptotic function at the mitochondrial level and caspases as the central player in the initiation and execution of apoptotic cell death have been the center of attraction for drug discovery studies and development of anticancer agents [10, 115, 116]. Here, various therapeutic strategies designed to target them have been reviewed.

2.3.1 Therapeutic opportunities based on Bcl-2 family proteins modulation

In view of the critical role of Bcl-2 proteins in regulation of mitochondrial pathway of apoptosis, targeting various members of this family have been considered amongst the most promising therapeutic strategies in cancer, a well-known dysfunctional apoptosis disorder [117]. Numerous attempts have been carried out to target the modifications in Bim expression and therefore regulate tumor cell response to apoptosis. Histone deacetylase inhibitors have been shown not only cause to up regulation of Bim in transformed cells, but also they are able to reverse silencing of Bim in cancer cells and consequently restored their sensitivity to various anticancer-agents reported in leukemia and Burkitt's lymphoma cells [118]. The proteasome inhibitors are also recognized to promote accumulation of Bim and enhance the lethality of cancer cells [119, 120]. Another approach is through diminishing its degradation by blocking its phosphorylation. Ras/Raf/MEK/ERK pathway have a key role in regulating the expression and function of Bim through its phosphorylation and triggering its proteasomal degradation. MEK1/2 Inhibitors has been applied to disrupt this process leading to accumulation of Bim and consequently apoptosis. MEK1/2 Inhibitors are also able to modify the interaction between BIM and other Bcl-2 family members contributing to cell death [118, 121, 122].

Furthermore, structure-based drug design can be applied to discover anticancer agents which are able to effectively activate a pro-apoptotic Bcl-2 protein through changing its conformation promoting cell death. Bax as a unique entry point for intrinsic apoptotic signaling is another major pro-apoptotic member of the Bcl-2 family proteins which has been greatly getting attention to be targeted in order to control apoptosis. Recent studies have revealed that direct binding and activation of Bax can be a promising approach for cancer treatment. Discovery of small-molecule functioning as a Bax activators may result in selective induction of tumor cell apoptosis and overcome chemoresistance which has been proved through invitro and invivo studies [117, 123]. Besides, some studies targeting a regulatory site in Ser184 of Bax protein have determined that its agonists SMBA1–SMBA3 can effectively bind to and trigger its oligomerization through the suppression of its phosphorylation that eventually lead to cyrochrome c release and induction of apoptosis in mouse lung cancer xenografts [124]. Similar results were also reported with other Bax agonists as promising Bax direct activators in breast cancer, glioblastoma and acute myeloid leukemia cells. These drug candidates demonstrated noteworthy in vivo efficiency inhibiting xenograft tumor growth though induction of apoptotic cell death [125–127].

The next emerging strategy in cancer drug discovery was the BH3 mimetics which are able to antagonize the function of Bcl-2 and selectively kill cancer cells. In this approach, BH3 mimetics are antagonists of the anti-apoptotic Bcl-2 proteins. These small molecules acting as the competitive inhibitors induce apoptosis though binding to their hydrophobic cleft and therefore affect the interactions between anti- and pro-apoptotic proteins [128]. Various BH3 mimetics with different level of specificity and efficiency have been reported. For instance, TW-37 derived from phenolic aldehyde gossypol has been showing high affinity to bind MCL-1, Bcl-2 and Bcl-xL anti-apoptotic proteins and induce apoptosis in B-cell lymphomas and pancreatic cell lines along with decreasing tumor size in xenograft models [129–131]. As ABT-737 mimicked the BH3 domain of Bad protein, it was able to bind selectively to Bcl-2, Bcl-xL and Bcl-W. It also demonstrated poor affinity to other member of ani-apoptic proteins including MCL-1 and BFL-1. ABT-737 has shown efficacy in the killing of several cancer cell lines including leukemia, lymphoma, multiple myeloma, glioma and small cell lung cancer cell lines as well as primary samples. Also, these two inhibitors of Bcl-2 families are currently in clinical trials [132–134].

Another approach to antagonize the function of Bcl-2 anti-apoptotic proteins is focusing on the protein interaction among members of Bcl-2 family through their essential death domain. In this regard, peptide-based inhibitors have been significant achievements in targeting and regulating intracellular protein–protein interaction. Stapled peptides are synthetic, bioactive α-helical peptides locked into their bioactive structure that have brought new hope to target drug discovery [135, 136]. For instance, stabilized alpha-helix of Bcl-2 domains, SAHBs, is the peptide having the ability to penetrate leukaemia cells and trigger induction of apoptosis through its binding to the Bcl-xL which its function has been further confirmed though invivo mouse xenograft models of leukaemia [137]. Another research study has also revealed that exclusive MCL-1 stapled peptide inhibitor (MCL-1 SAHBD) can effectively resensitize cancer cells to caspase-mediated apoptosis through directly targeting of MCL-1 and suppress its inhibitory interaction with Bak protein [138].

2.3.2 Therapeutic opportunities based on caspase modulation

Given the vital role of caspases in the regulation of apoptosis, it is not surprising that numerous therapeutic opportunities targeting caspase activity demonstrate great promise for the cancer treatment. Different strategies have been investigated to upregulate caspase-8 expression to restore its function in tumors. As hypermethylation of its promotor has been recognized as the main mechanism of silencing, one approach for its reactivation is using demethylation agents. Azacytidine, decitabine and nucleoside analogs promoting the demethylation of caspase-8 promotor have been successfully applied in neuroblastoma, medulloblastoma, breast cancer and lung carcinoma [139, 140]. Another interesting strategy is designing the small molecules that selectively and directly target and trigger caspase-8 activation. These small molecules has been reported to potentiate TRAIL-induced cell death [141]. Proteosomal inhibitors such as bortezomib has been also reported to increase total cellular caspase-8 levels apparently by blocking its degradation [111, 142]. Some studies have also reported that the use of interferons can elevate the caspase-8 expression through modification at transcriptional level. This strategy targeting interferon-sensitive response elements within the caspase-8 promoter leading to sensitize cancer cell to apoptotic cell death in cancer chemotherapy or irradiation therapy [139, 143, 144].

Besides, developing molecules that are able to directly activates caspase 3 have been of research interests as well. For this purpose, particular sequence of inactive procaspase-3 consisting of the triplet of aspartic acid residues has been targeted. In vitro studies have exhibited that PETCM, gambonic acid and its derivatives have the potential to effectively activate caspase 3 leading to apoptotic cell death in cancer cell lines [145–147]. Furthermore, procaspase-activating compound1 (PAC-1) has been shown to induce anticancer activity through promoting the procaspase-3 activation. PAC-1 exerted its effect by chelation of inhibitory labile zinc ions and currently is in phase I clinical trial for cancer treatment [148].

In order to sensitize tumor cells to apoptotic stimuli, caspase −9 can be also regarded as a potential target in cancer therapy. There are a wide range of molecules such as protein kinase, microRNAs and heat shock protein that have been identified to modulate caspase-9 expression and hence have been getting interest as candidates for new drug development though regulating intrinsic apoptosis in cancer cells [149, 150]. Targeting caspase-9 have been also initiated in clinical trials (phase I) against several cancer including Chronic Myeloid Leukemia, non-Hodgkin's lymphoma, Acute Lymphoblastic Leukemia, [151, 152].

In addition, several attempts have also been conducted on cancer gene therapy focusing on apoptotic caspases. Gene transfer technologies may restore caspase gene expression resulting in selectively induction of apoptosis in various tumor types [153–155]. In this regard, caspase-9 and caspase-3 has been suggested for being used in gene therapy strategies. A main benefit of involving these caspases is that they start apoptosis at the downstream of the mitochondrial outer membrane potential and they will not be affected with the enhanced expression of anti-apoptotic of Bcl-2 proteins. The researchers conducted on inducible version of these caspases have shown encouraging results related to remarkable reduction in size of lung and gastric tumors, respectively [156–158].

Other than directly targeting of caspases, another area of research has focused on discovery of anticancer agents that trigger the caspases activity indirectly. In this approach, certain members of the inhibitors of apoptosis proteins (IAP) are targeted. IAPs are functioning as the endogenous caspase inhibitors and prevent apoptosis event by binding and inhibiting caspases through the degradation of active caspases or keeping them away from their substrate. In this regard, numerous researches have investigated various IAP inhibiting agents, accomplishing a breakthrough in cancer treatment [159, 160]. Some of these agents are acting as the IAP antagonist and exert their effect via suppression of their activity, while others are analogs of the endogenous IAP inhibitor Smac. Several Smac mimetics such as LCL161 and birinapant IAP inhibitors have currently being tested in phaseI/II in clinical trials, with promising outcomes [161–164]. Besides, IAP inhibitors have been reported to exert the synergistic effect in combination chemotherapy and sensitize the cancer cells to radiotherapy which is of particular interest in malignant gliomas [165–167].

3. Conclusion

It is well established that the apoptosis dysfunction promotes the malignant transformation and renders the cancer cell resistant to treatment. Targeting apoptotic pathways in tumor cells has been a main clinical interest as the evasion of apoptosis is a hallmark of all cancers regardless of their causes or types. There are numerous defects found in apoptotic mechanism contributing to inhibition of cancer cell death. As demonstrated in this chapter, impaired activation of caspases and disturbance in the balance between anti-apoptotic and pro-apoptotic members of Bcl-2 family proteins are remarkably involved in tumorgenesis. The enhanced knowledge about their critical roles in apoptosis and cell fate in recent years has eventually made them promising therapeutic targets. This also has facilitated the generation of more specific anticancer agents and led to shifting in anticancer therapy form typical cytotoxic approaches to the designing and development of apoptosis-inducing drugs that particularly target the cancer cells. An exciting development in successful eradication of cancer cells involves structure-based drug design of small molecules such as BH3 mimetics, specifically targeting Bcl-2 proteins, that is currently being tested in clinical trials with promising effects of selective induction of tumor cell apoptosis and overcoming chemoresistance as well. These inhibitor molecules are in continuous development and a great deal of effort is required to discover the most efficient ones having more specificity for individual Bcl-2 proteins and offer maximal clinical efficacy. Besides, new therapeutic applications targeting apoptotic caspases including gene therapy approaches and small molecules suppressing inhibitors of caspases are beginning to show some promise through selectively and directly targeting of individual caspases and eventually triggering their activity. Caspase-targeted approaches, epigenetic modulators and their combinations with established therapies may have the potential to overcome the limitation of previous strategies through exerting synergistic pro-apoptotic activity and may enhance the effectiveness of conventional cancer therapy, worthy of further investigation in preclinical advanced models and clinical trial. Apoptosis-targeted therapies are now remarkably advancing and remain a promising approaches in future clinical practice of oncology.

Acknowledgements

I would like to thank Prof. Mahmood Ameen Abdulla and Dr. Jayakumar Rajarajeswaran for their valuable guidance.

Conflict of interest

There is no conflict of interest.

Author details

Reyhaneh Farghadani* and Rakesh Naidu*
Jeffrey Cheah School of Medicine and Health Sciences, Monash University Malaysia, Bandar Sunway, Selangor Darul Ehsan, Malaysia

*Address all correspondence to: r_farghadani@yahoo.com; reyhaneh.farghadani@monash.edu and rakesh.naidu@monash.edu

References

[1] Hejmadi M. Introduction to cancer biology. London: Bookboon; 2009.

[2] World Health Organization. Cancer 2018 [Available from: http://www.who.int/news-room/fact-sheets/detail/cancer.

[3] Bray F, Ferlay J, Soerjomataram I, Siegel RL, Torre LA, Jemal A. Global cancer statistics 2018: GLOBOCAN estimates of incidence and mortality worldwide for 36 cancers in 185 countries. CA Cancer J Clin. 2018;68(6):394-424.

[4] Ferlay J, Soerjomataram I, Dikshit R, Eser S, Mathers C, Rebelo M, et al. Cancer incidence and mortality worldwide: sources, methods and major patterns in GLOBOCAN 2012. Int J Cancer. 2015;136(5):E359-E86.

[5] Siegel RL, Miller KD, Jemal A. Cancer statistics, 2015. CA Cancer J Clin. 2015;65(1):5-29.

[6] Fernald K, Kurokawa M. Evading apoptosis in cancer. Trends Cell Biol. 2013;23(12):620-33.

[7] Pfeffer C, Singh A. Apoptosis: a target for anticancer therapy. Int J Mol Sci. 2018;19(2):448.

[8] Carneiro BA, El-Deiry WS. Targeting apoptosis in cancer therapy. Nature reviews Clinical oncology. 2020;17(7):395-417.

[9] Campbell KJ, Tait SW. Targeting BCL-2 regulated apoptosis in cancer. Open biology. 2018;8(5):180002.

[10] Pistritto G, Trisciuoglio D, Ceci C, Garufi A, D'Orazi G. Apoptosis as anticancer mechanism: function and dysfunction of its modulators and targeted therapeutic strategies. Aging (Albany NY). 2016;8(4):603.

[11] Koff JL, Ramachandiran S, Bernal-Mizrachi L. A time to kill: targeting apoptosis in cancer. Int J Mol Sci. 2015;16(2):2942-55.

[12] Wong RS. Apoptosis in cancer: from pathogenesis to treatment. J Exp Clin Cancer Res. 2011;30(1):87.

[13] Elmore S. Apoptosis: a review of programmed cell death. Toxicol Pathol. 2007;35(4):495-516.

[14] Kerr JF, Wyllie AH, Currie AR. Apoptosis: a basic biological phenomenon with wideranging implications in tissue kinetics. Br J Cancer. 1972;26(4):239.

[15] Pfeffer CM, Singh AT. Apoptosis: a target for anticancer therapy. Int J Mol Sci. 2018;19(2):448.

[16] Goodsell DS. The molecular perspective: caspases. Stem Cells. 2000;18(6):457-8.

[17] McIlwain DR, Berger T, Mak TW. Caspase functions in cell death and disease. Cold Spring Harb Perspect Biol. 2015;7(4).

[18] Parrish AB, Freel CD, Kornbluth S. Cellular mechanisms controlling caspase activation and function. Cold Spring Harb Perspect Biol. 2013;5(6):a008672.

[19] Li J, Yuan J. Caspases in apoptosis and beyond. Oncogene. 2008;27(48):6194.

[20] Shi Y. Caspase activation: revisiting the induced proximity model. Cell. 2004;117(7):855-8.

[21] Cohen GM. Caspases: the executioners of apoptosis. Biochem J. 1997;326(1):1-16.

[22] Kitazumi I, Tsukahara M. Regulation of DNA fragmentation: the

role of caspases and phosphorylation. The FEBS journal. 2011;278(3):427-41.

[23] Doonan F, Cotter TG. Morphological assessment of apoptosis. Methods. 2008;44(3):200-4.

[24] Farghadani R, Rajarajeswaran J, Hashim NBM, Abdulla MA, Muniandy S. A novel β-diiminato manganese III complex as the promising anticancer agent induces G 0/G 1 cell cycle arrest and triggers apoptosis via mitochondrial-dependent pathways in MCF-7 and MDA-MB-231 human breast cancer cells. RSC Advances. 2017;7(39):24387-98.

[25] Caruso S, Atkin-Smith GK, Baxter AA, Tixeira R, Jiang L, Ozkocak DC, et al. Defining the role of cytoskeletal components in the formation of apoptopodia and apoptotic bodies during apoptosis. Apoptosis. 2019;24(11-12):862-77.

[26] Model MA, Schonbrun E. Optical determination of intracellular water in apoptotic cells. The Journal of Physiology. 2013;591(23):5843-9.

[27] Gómez-Angelats M, Cidlowski JA. Invited review: cell volume control and signal transduction in apoptosis. Toxicol Pathol. 2002;30(5):541-51.

[28] Orlov SN, Platonova AA, Hamet P, Grygorczyk R. Cell volume and monovalent ion transporters: their role in cell death machinery triggering and progression. American Journal of Physiology-Cell Physiology. 2013.

[29] Julian L, Olson MF. Apoptotic membrane dynamics in health and disease. Cell Health and Cytoskeleton. 2015;2015(7):133-42.

[30] Segawa K, Nagata S. An Apoptotic 'Eat Me' Signal: Phosphatidylserine Exposure. Trends Cell Biol. 2015;25(11):639-50.

[31] Fadok VA, de Cathelineau A, Daleke DL, Henson PM, Bratton DL. Loss of phospholipid asymmetry and surface exposure of phosphatidylserine is required for phagocytosis of apoptotic cells by macrophages and fibroblasts. J Biol Chem. 2000.

[32] Green DR, Llambi F. Cell death signaling. Cold Spring Harb Perspect Biol. 2015;7(12):a006080.

[33] Igney FH, Krammer PH. Death and anti-death: tumour resistance to apoptosis. Nature Reviews Cancer. 2002;2(4):277.

[34] Li MX, Dewson G. Mitochondria and apoptosis: emerging concepts. F1000prime reports. 2015;7.

[35] Wu C-C, Bratton SB. Regulation of the intrinsic apoptosis pathway by reactive oxygen species. Antioxidants & redox signaling. 2013;19(6):546-58.

[36] Montero J, Letai A. Why do BCL-2 inhibitors work and where should we use them in the clinic? Cell Death Differ. 2018;25(1):56.

[37] Shamas-Din A, Kale J, Leber B, Andrews DW. Mechanisms of action of Bcl-2 family proteins. Cold Spring Harb Perspect Biol. 2013;5(4):a008714.

[38] Siddiqui WA, Ahad A, Ahsan H. The mystery of BCL2 family: Bcl-2 proteins and apoptosis: an update. Arch Toxicol. 2015;89(3):289-317.

[39] Chipuk JE, Green DR. How do BCL-2 proteins induce mitochondrial outer membrane permeabilization? Trends Cell Biol. 2008;18(4):157-64.

[40] Tait SW, Green DR. Mitochondria and cell death: outer membrane permeabilization and beyond. Nat Rev Mol Cell Biol. 2010;11(9):621-32.

[41] Slee EA, Harte MT, Kluck RM, Wolf BB, Casiano CA, Newmeyer DD,

et al. Ordering the cytochrome c–initiated caspase cascade: hierarchical activation of caspases-2,-3,-6,-7,-8, and-10 in a caspase-9–dependent manner. The Journal of cell biology. 1999;144(2):281-92.

[42] Nair P, Lu M, Petersen S, Ashkenazi A. Apoptosis initiation through the cell-extrinsic pathway. Methods Enzymol. 544: Elsevier; 2014. p. 99-128.

[43] Ricci MS, El-Deiry WS. The extrinsic pathway of apoptosis. Apoptosis, Senescence, and Cancer: Springer; 2007. p. 31-54.

[44] Behzadi P, Ranjbar R. Caspases and apoptosis. Cell. 2015;4(7):11-3.

[45] Jänicke RU, Sprengart ML, Wati MR, Porter AG. Caspase-3 is required for DNA fragmentation and morphological changes associated with apoptosis. J Biol Chem. 1998;273(16):9357-60.

[46] Fischer U, Schulze-Osthoff K. Apoptosis-based therapies and drug targets. Cell Death Differ. 2005;12(S1):942.

[47] Hatok J, Racay P. Bcl-2 family proteins: master regulators of cell survival. Biomolecular Concepts. 2016;7(4):259-70.

[48] Gross A, McDonnell JM, Korsmeyer SJ. BCL-2 family members and the mitochondria in apoptosis. Genes Dev. 1999;13(15):1899-911.

[49] Shamas-Din A, Brahmbhatt H, Leber B, Andrews DW. BH3-only proteins: Orchestrators of apoptosis. Biochimica et Biophysica Acta (BBA)-Molecular Cell Research. 2011;1813(4):508-20.

[50] Dewson G, Kluck RM. Bcl-2 family-regulated apoptosis in health and disease. Cell Health and Cytoskeleton. 2010;2:9-22.

[51] Petros AM, Olejniczak ET, Fesik SW. Structural biology of the Bcl-2 family of proteins. Biochimica et Biophysica Acta (BBA)-Molecular Cell Research. 2004;1644(2-3):83-94.

[52] Amundson SA, Myers TG, Scudiero D, Kitada S, Reed JC, Fornace AJ. An informatics approach identifying markers of chemosensitivity in human cancer cell lines. Cancer Res. 2000;60(21):6101-10.

[53] Reed JC. Bcl-2–family proteins and hematologic malignancies: history and future prospects. Blood, The Journal of the American Society of Hematology. 2008;111(7):3322-30.

[54] Trisciuoglio D, Tupone MG, Desideri M, Di Martile M, Gabellini C, Buglioni S, et al. BCL-X L overexpression promotes tumor progression-associated properties. Cell Death Dis. 2017;8(12):1-15.

[55] D'Aguanno S, Del Bufalo D. Inhibition of anti-apoptotic Bcl-2 proteins in preclinical and clinical studies: current overview in cancer. Cells. 2020;9(5):1287.

[56] Li S, Li J, Hu T, Zhang C, Lv X, He S, et al. Bcl-2 overexpression contributes to laryngeal carcinoma cell survival by forming a complex with Hsp90β. Oncol Rep. 2017;37(2):849-56.

[57] Deng Y, Lin C, Zheng J, Fu M, Liang X, Chen J, et al. Overexpression of Bcl-2 partly inhibits apoptosis of human cervical cancer SiHa cells induced by arsenic trioxide. Chin Med J. 2000;113(1):84-8.

[58] Perini GF, Ribeiro GN, Neto JVP, Campos LT, Hamerschlak N. BCL-2 as therapeutic target for hematological malignancies. J Hematol Oncol. 2018;11(1):65.

[59] Huang J, Fairbrother W, Reed JC. Therapeutic targeting of Bcl-2 family

for treatment of B-cell malignancies. Expert Rev Hematol. 2015;8(3):283-97.

[60] Cimmino A, Calin GA, Fabbri M, Iorio MV, Ferracin M, Shimizu M, et al. miR-15 and miR-16 induce apoptosis by targeting BCL2. Proceedings of the National Academy of Sciences. 2005;102(39):13944-9.

[61] Schott A, Apel I, Nunez G, Clarke M. Bcl-XL protects cancer cells from p53-mediated apoptosis. Oncogene. 1995;11(7):1389-94.

[62] Jin-Song Y, Zhao-Xia W, Cheng-Yu L, Xiao-Di L, Ming S, Yuan-Yuan G, et al. Prognostic significance of Bcl-xL gene expression in human colorectal cancer. Acta Histochem. 2011;113(8):810-4.

[63] Scherr A-L, Gdynia G, Salou M, Radhakrishnan P, Duglova K, Heller A, et al. Bcl-x L is an oncogenic driver in colorectal cancer. Cell Death Dis. 2016;7(8):e2342-e.

[64] Watanabe J, Kushihata F, Honda K, Mominoki K, Matsuda S, Kobayashi N. Bcl-xL overexpression in human hepatocellular carcinoma. Int J Oncol. 2002;21(3):515-9.

[65] Castilla C, Congregado Bn, Chinchón D, Torrubia FJ, Japón MA, Sáez C. Bcl-xL is overexpressed in hormone-resistant prostate cancer and promotes survival of LNCaP cells via interaction with proapoptotic Bak. Endocrinology. 2006;147(10):4960-7.

[66] Chen J, Zhang X, Lentz C, Abi-Daoud M, Paré GC, Yang X, et al. miR-193b regulates Mcl-1 in melanoma. The American journal of pathology. 2011;179(5):2162-8.

[67] Gong J, Zhang J, Li B, Zeng C, You K, Chen M, et al. MicroRNA-125b promotes apoptosis by regulating the expression of Mcl-1, Bcl-w and IL-6R. Oncogene. 2013;32(25):3071-9.

[68] Shimizu S, Takehara T, Hikita H, Kodama T, Miyagi T, Hosui A, et al. The let-7 family of microRNAs inhibits Bcl-xL expression and potentiates sorafenib-induced apoptosis in human hepatocellular carcinoma. J Hepatol. 2010;52(5):698-704.

[69] Adams JM, Cory S. The Bcl-2 apoptotic switch in cancer development and therapy. Oncogene. 2007;26(9):1324-37.

[70] Konopleva M, Zhao S, Hu W, Jiang S, Snell V, Weidner D, et al. The anti-apoptotic genes Bcl-X(L) and Bcl-2 are over-expressed and contribute to chemoresistance of non-proliferating leukaemic CD34+ cells. Br J Haematol. 2002;118(2):521-34.

[71] de Melo Silva AJ. Bcl-2 Family Overexpression and Chemoresistance in Acute Myeloid Leukemia. Serbian Journal of Experimental and Clinical Research. 2018;19(4):299-309.

[72] Maji S, Panda S, Samal SK, Shriwas O, Rath R, Pellecchia M, et al. Bcl-2 antiapoptotic family proteins and chemoresistance in cancer. Adv Cancer Res. 137: Elsevier; 2018. p. 37-75.

[73] Anderton E, Yee J, Smith P, Crook T, White R, Allday M. Two Epstein–Barr virus (EBV) oncoproteins cooperate to repress expression of the proapoptotic tumour-suppressor Bim: clues to the pathogenesis of Burkitt's lymphoma. Oncogene. 2008;27(4):421-33.

[74] Richter-Larrea JA, Robles EF, Fresquet V, Beltran E, Rullan AJ, Agirre X, et al. Reversion of epigenetically mediated BIM silencing overcomes chemoresistance in Burkitt lymphoma. Blood, The Journal of the American Society of Hematology. 2010;116(14):2531-42.

[75] Tagawa H, Karnan S, Suzuki R, Matsuo K, Zhang X, Ota A, et al. Genome-wide array-based CGH for

mantle cell lymphoma: identification of homozygous deletions of the proapoptotic gene BIM. Oncogene. 2005;24(8):1348-58.

[76] Zantl N, Weirich G, Zall H, Seiffert BM, Fischer SF, Kirschnek S, et al. Frequent loss of expression of the pro-apoptotic protein Bim in renal cell carcinoma: evidence for contribution to apoptosis resistance. Oncogene. 2007;26(49):7038-48.

[77] Manoochehri M, Karbasi A, Bandehpour M, Kazemi B. Down-regulation of BAX gene during carcinogenesis and acquisition of resistance to 5-FU in colorectal cancer. Pathology & Oncology Research. 2014;20(2):301-7.

[78] Perego P, Giarola M, Righetti SC, Supino R, Caserini C, Delia D, et al. Association between cisplatin resistance and mutation of p53 gene and reduced bax expression in ovarian carcinoma cell systems. Cancer Res. 1996;56(3):556-62.

[79] Aoyagi T, Morii T, Ohtsuka K, Ohnishi H, Tajima T, Yoshiyama A, et al. Lung cancer cell line sensitivity to zoledronic acid is BAX-dependent. Anticancer Res. 2013;33(12):5357-63.

[80] Ravi R, Bedi A. Requirement of BAX for TRAIL/Apo2L-induced apoptosis of colorectal cancers: synergism with sulindac-mediated inhibition of Bcl-xL. Cancer Res. 2002;62(6):1583-7.

[81] Eberle J. Countering trail resistance in melanoma. Cancers (Basel). 2019;11(5):656.

[82] Deng Y, Lin Y, Wu X. TRAIL-induced apoptosis requires Bax-dependent mitochondrial release of Smac/DIABLO. Genes Dev. 2002;16(1):33-45.

[83] Miquel C, Borrini F, Grandjouan S, Aupérin A, Viguier J, Velasco V, et al. Role of bax mutations in apoptosis in colorectal cancers with microsatellite instability. Am J Clin Pathol. 2005;123(4):562-70.

[84] Fresquet V, Rieger M, Carolis C, García-Barchino MJ, Martinez-Climent JA. Acquired mutations in BCL2 family proteins conferring resistance to the BH3 mimetic ABT-199 in lymphoma. Blood. 2014;123(26):4111-9.

[85] Kitada S, Pedersen IM, Schimmer AD, Reed JC. Dysregulation of apoptosis genes in hematopoietic malignancies. Oncogene. 2002;21(21):3459-74.

[86] Wei MC, Zong W-X, Cheng EH-Y, Lindsten T, Panoutsakopoulou V, Ross AJ, et al. Proapoptotic BAX and BAK: a requisite gateway to mitochondrial dysfunction and death. Science. 2001;292(5517):727-30.

[87] Fiandalo M, Kyprianou N. Caspase control: protagonists of cancer cell apoptosis. Exp Oncol. 2012;34(3):165.

[88] Ando M, Kawazu M, Ueno T, Fukumura K, Yamato A, Soda M, et al. Cancer-associated missense mutations of caspase-8 activate nuclear factor-κB signaling. Cancer Sci. 2013;104(8):1002-8.

[89] Teng Y, Dong YC, Liu Z, Zou Y, Xie H, Zhao Y, et al. DNA methylation-mediated caspase-8 downregulation is associated with anti-apoptotic activity and human malignant glioma grade. Int J Mol Med. 2017;39(3):725-33.

[90] Shen XG, Wang C, Li Y, Wang L, Zhou B, Xu B, et al. Downregulation of caspase-9 is a frequent event in patients with stage II colorectal cancer and correlates with poor clinical outcome. Colorectal Dis. 2010;12(12):1213-8.

[91] Jing Z, Heng W, Xia L, Ning W, Yafei Q, Yao Z, et al. Downregulation of

phosphoglycerate dehydrogenase inhibits proliferation and enhances cisplatin sensitivity in cervical adenocarcinoma cells by regulating Bcl-2 and caspase-3. Cancer Biol Ther. 2015;16(4):541-8.

[92] Zheng Ba, Chai R, Yu X. Downregulation of NIT2 inhibits colon cancer cell proliferation and induces cell cycle arrest through the caspase-3 and PARP pathways. Int J Mol Med. 2015;35(5):1317-22.

[93] Jia G, Wang Q, Wang R, Deng D, Xue L, Shao N, et al. Tubeimoside-1 induces glioma apoptosis through regulation of Bax/Bcl-2 and the ROS/Cytochrome C/Caspase-3 pathway. Onco Targets Ther. 2015;8:303.

[94] Zhou L, Luo Y, Li K, Tian L, Wang M, Li C, et al. Molecular markers of therapeutic resistance in breast cancer. Hum Pathol. 2013;44(7):1421-8.

[95] Zhou M, Liu X, Li Z, Huang Q, Li F, Li CY. Caspase-3 regulates the migration, invasion and metastasis of colon cancer cells. Int J Cancer. 2018;143(4):921-30.

[96] Pu X, Storr SJ, Zhang Y, Rakha EA, Green AR, Ellis IO, et al. Caspase-3 and caspase-8 expression in breast cancer: caspase-3 is associated with survival. Apoptosis. 2017;22(3):357-68.

[97] Palmerini F, Devilard E, Jarry A, Birg F, Xerri L. Caspase 7 downregulation as an immunohistochemical marker of colonic carcinoma. Hum Pathol. 2001;32(5):461-7.

[98] Yang X, Zhong D-N, Qin H, Wu P-R, Wei K-L, Chen G, et al. Caspase-3 over-expression is associated with poor overall survival and clinicopathological parameters in breast cancer: a meta-analysis of 3091 cases. Oncotarget. 2018;9(9):8629.

[99] Zhao R, Kaakati R, Lee AK, Liu X, Li F, Li C-Y. Novel roles of apoptotic caspases in tumor repopulation, epigenetic reprogramming, carcinogenesis, and beyond. Cancer Metastasis Rev. 2018;37(2-3):227-36.

[100] Kumar S. Caspase function in programmed cell death. Cell Death Differ. 2007;14(1):32-43.

[101] Asadi M, Shanehbandi D, Kermani TA, Sanaat Z, Zafari V, Hashemzadeh S. Expression level of caspase genes in colorectal cancer. Asian Pacific journal of cancer prevention: APJCP. 2018;19(5):1277.

[102] Kuwahara D, Tsutsumi K, Oyake D, Ohta T, Nishikawa H, Koizuka I. Inhibition of caspase-9 activity and Apaf-1 expression in cisplatin-resistant head and neck squamous cell carcinoma cells. Auris Nasus Larynx. 2003;30:85-8.

[103] Liamarkopoulos E, Gazouli M, Aravantinos G, Tzanakis N, Theodoropoulos G, Rizos S, et al. Caspase 8 and caspase 9 gene polymorphisms and susceptibility to gastric cancer. Gastric Cancer. 2011;14(4):317-21.

[104] Li P, Zhou L, Zhao T, Liu X, Zhang P, Liu Y, et al. Caspase-9: structure, mechanisms and clinical application. Oncotarget. 2017;8(14):23996.

[105] Yilmaz SG, Yencilek F, Yildirim A, Yencilek E, Isbir T. Effects of caspase 9 gene polymorphism in patients with prostate cancer. In Vivo. 2017;31(2):205-8.

[106] Ercan S, Arinc S, Yilmaz SG, Altunok C, Yaman F, Isbir T. Investigation of Caspase 9 Gene Polymorphism in Patients With Non-small Cell Lung Cancer. Anticancer Res. 2019;39(5):2437-41.

[107] Xu B, Zhou Z-G, Li Y, Wang L, Yang L, Zhou B, et al.

Clinicopathological significance of caspase-8 and caspase-10 expression in rectal cancer. Oncology. 2008;74(3-4):229-36.

[108] Fong P, Xue W, Ngan H, Chiu P, Chan K, Tsao S, et al. Caspase activity is downregulated in choriocarcinoma: a cDNA array differential expression study. J Clin Pathol. 2006;59(2):179-83.

[109] Aghababazadeh M, Dorraki N, Javan FA, Fattahi AS, Gharib M, Pasdar A. Downregulation of Caspase 8 in a group of Iranian breast cancer patients–A pilot study. J Egypt Natl Canc Inst. 2017;29(4):191-5.

[110] Kostova I, Mandal R, Becker S, Strebhardt K. The role of caspase-8 in the tumor microenvironment of ovarian cancer. Cancer Metastasis Rev. 2020:1-16.

[111] Stupack DG. Caspase-8 as a therapeutic target in cancer. Cancer Lett. 2013;332(2):133-40.

[112] Graf RP, Keller N, Barbero S, Stupack D. Caspase-8 as a regulator of tumor cell motility. Curr Mol Med. 2014;14(2):246-54.

[113] Pingoud-Meier C, Lang D, Janss AJ, Rorke LB, Phillips PC, Shalaby T, et al. Loss of caspase-8 protein expression correlates with unfavorable survival outcome in childhood medulloblastoma. Clin Cancer Res. 2003;9(17):6401-9.

[114] Elrod HA, Fan S, Muller S, Chen GZ, Pan L, Tighiouart M, et al. Analysis of death receptor 5 and caspase-8 expression in primary and metastatic head and neck squamous cell carcinoma and their prognostic impact. PLoS One. 2010;5(8):e12178.

[115] Baig S, Seevasant I, Mohamad J, Mukheem A, Huri H, Kamarul T. Potential of apoptotic pathway-targeted cancer therapeutic research: Where do we stand? Cell Death Dis. 2016;7(1):e2058-e.

[116] Boice A, Bouchier-Hayes L. Targeting apoptotic caspases in cancer. Biochimica et Biophysica Acta (BBA)-Molecular Cell Research. 2020;1867(6):118688.

[117] Liu Z, Ding Y, Ye N, Wild C, Chen H, Zhou J. Direct activation of bax protein for cancer therapy. Med Res Rev. 2016;36(2):313-41.

[118] Harada H, Grant S. Targeting the regulatory machinery of BIM for cancer therapy. Critical Reviews™ in Eukaryotic Gene Expression. 2012;22(2).

[119] Pigneux A, Mahon F-X, Moreau-Gaudry F, Uhalde M, De Verneuil H, Lacombe F, et al. Proteasome inhibition specifically sensitizes leukemic cells to anthracyclin-induced apoptosis through the accumulation of Bim and Bax pro-apoptotic proteins. Cancer Biol Ther. 2007;6(4):603-11.

[120] Dai Y, Chen S, Wang L, Pei X-Y, Kramer LB, Dent P, et al. Bortezomib interacts synergistically with belinostat in human aml and all cells in association with perturbations in NF-κB and BIM. Br J Haematol. 2011;153(2):222.

[121] Ley R, Balmanno K, Hadfield K, Weston C, Cook SJ. Activation of the ERK1/2 signaling pathway promotes phosphorylation and proteasome-dependent degradation of the BH3-only protein, Bim. J Biol Chem. 2003;278(21):18811-6.

[122] Ewings KE, Wiggins CM, Cook SJ. Bim and the pro-survival Bcl-2 proteins: opposites attract, ERK repels. Cell cycle. 2007;6(18):2236-40.

[123] Zhao G, Zhu Y, Eno CO, Liu Y, DeLeeuw L, Burlison JA, et al. Activation of the proapoptotic Bcl-2 protein Bax by a small molecule induces tumor cell apoptosis. Mol Cell Biol. 2014;34(7):1198-207.

[124] Xin M, Li R, Xie M, Park D, Owonikoko TK, Sica GL, et al. Small-molecule Bax agonists for cancer therapy. Nature communications. 2014;5(1):1-11.

[125] Liu G, Kim H, Ding C, Yu Z, Wang H, Chen H, et al. Abstract 3: Discovery and optimization of small molecule Bax activators for cancer therapy. Cancer Res. 2019;79(13 Supplement):3-.

[126] Daniele S, Pietrobono D, Costa B, Giustiniano M, La Pietra V, Giacomelli C, et al. Bax activation blocks self-renewal and induces apoptosis of human glioblastoma stem cells. ACS Chem Neurosci. 2018;9(1):85-99.

[127] Reyna DE, Garner TP, Lopez A, Kopp F, Choudhary GS, Sridharan A, et al. Direct activation of BAX by BTSA1 overcomes apoptosis resistance in acute myeloid leukemia. Cancer Cell. 2017;32(4):490-505. e10.

[128] Chonghaile TN, Letai A. Mimicking the BH3 domain to kill cancer cells. Oncogene. 2008;27(1):S149-S57.

[129] Verhaegen M, Bauer JA, de la Vega CM, Wang G, Wolter KG, Brenner JC, et al. A novel BH3 mimetic reveals a mitogen-activated protein kinase–dependent mechanism of melanoma cell death controlled by p53 and reactive oxygen species. Cancer Res. 2006;66(23):11348-59.

[130] Wang Z, Azmi AS, Ahmad A, Banerjee S, Wang S, Sarkar FH, et al. TW-37, a small-molecule inhibitor of Bcl-2, inhibits cell growth and induces apoptosis in pancreatic cancer: involvement of Notch-1 signaling pathway. Cancer Res. 2009;69(7):2757-65.

[131] Mohammad RM, Goustin AS, Aboukameel A, Chen B, Banerjee S, Wang G, et al. Preclinical studies of TW-37, a new nonpeptidic small-molecule inhibitor of Bcl-2, in diffuse large cell lymphoma xenograft model reveal drug action on both Bcl-2 and Mcl-1. Clin Cancer Res. 2007;13(7):2226-35.

[132] Chauhan D, Velankar M, Brahmandam M, Hideshima T, Podar K, Richardson P, et al. A novel Bcl-2/Bcl-X L/Bcl-w inhibitor ABT-737 as therapy in multiple myeloma. Oncogene. 2007;26(16):2374-80.

[133] Tagscherer KE, Fassl A, Campos B, Farhadi M, Kraemer A, Böck B, et al. Apoptosis-based treatment of glioblastomas with ABT-737, a novel small molecule inhibitor of Bcl-2 family proteins. Oncogene. 2008;27(52):6646-56.

[134] Xia J, Bai H, Yan B, Li R, Shao M, Xiong L, et al. Mimicking the BIM BH3 domain overcomes resistance to EGFR tyrosine kinase inhibitors in EGFR-mutant non-small cell lung cancer. Oncotarget. 2017;8(65):108522.

[135] Ali AM, Atmaj J, Van Oosterwijk N, Groves MR, Dömling A. Stapled peptides inhibitors: a new window for target drug discovery. Computational and structural biotechnology journal. 2019;17:263-81.

[136] McDougall L, Jamieson AG. Stapled Peptides as Potential Therapeutics. eLS. 2019:1-9.

[137] Walensky LD, Kung AL, Escher I, Malia TJ, Barbuto S, Wright RD, et al. Activation of apoptosis in vivo by a hydrocarbon-stapled BH3 helix. Science. 2004;305(5689):1466-70.

[138] Stewart ML, Fire E, Keating AE, Walensky LD. The MCL-1 BH3 helix is an exclusive MCL-1 inhibitor and apoptosis sensitizer. Nat Chem Biol. 2010;6(8):595-601.

[139] Fulda S, editor Therapeutic opportunities based on caspase modulation. Semin Cell Dev Biol; 2018: Elsevier.

[140] Hensley P, Mishra M, Kyprianou N. Targeting caspases in cancer therapeutics. Biol Chem. 2013;394(7):831-43.

[141] Bucur O, Gaidos G, Yatawara A, Pennarun B, Rupasinghe C, Roux J, et al. A novel caspase 8 selective small molecule potentiates TRAIL-induced cell death. Sci Rep. 2015;5:9893.

[142] Thorpe JA, Christian PA, Schwarze SR. Proteasome inhibition blocks caspase-8 degradation and sensitizes prostate cancer cells to death receptor-mediated apoptosis. The Prostate. 2008;68(2):200-9.

[143] Häcker S, Dittrich A, Mohr A, Schweitzer T, Rutkowski S, Krauss J, et al. Histone deacetylase inhibitors cooperate with IFN-γ to restore caspase-8 expression and overcome TRAIL resistance in cancers with silencing of caspase-8. Oncogene. 2009;28(35):3097.

[144] Ruiz-Ruiz C, de Almodóvar CR, Rodríguez A, Ortiz-Ferrón G, Redondo JM, López-Rivas A. The up-regulation of human caspase-8 by interferon-γ in breast tumor cells requires the induction and action of the transcription factor interferon regulatory factor-1. J Biol Chem. 2004;279(19):19712-20.

[145] Zhang H-Z, Kasibhatla S, Wang Y, Herich J, Guastella J, Tseng B, et al. Discovery, characterization and SAR of gambogic acid as a potent apoptosis inducer by a HTS assay. Biorg Med Chem. 2004;12(2):309-17.

[146] Roy S, Bayly CI, Gareau Y, Houtzager VM, Kargman S, Keen SL, et al. Maintenance of caspase-3 proenzyme dormancy by an intrinsic “safety catch” regulatory tripeptide. Proceedings of the National Academy of Sciences. 2001;98(11):6132-7.

[147] McIlwain DR, Berger T, Mak TW. Caspase functions in cell death and disease. Cold Spring Harb Perspect Biol. 2013;5(4):a008656.

[148] S Roth H, J Hergenrother P. Derivatives of procaspase-activating compound 1 (PAC-1) and their anticancer activities. Curr Med Chem. 2016;23(3):201-41.

[149] Kim B, Srivastava SK, Kim S-H. Caspase-9 as a therapeutic target for treating cancer. Expert Opin Ther Targets. 2015;19(1):113-27.

[150] Serrano BP, Hardy JA. Phosphorylation by protein kinase A disassembles the caspase-9 core. Cell Death Differ. 2018;25(6):1025-39.

[151] Zhou X, Dotti G, Krance RA, Martinez CA, Naik S, Kamble RT, et al. Inducible caspase-9 suicide gene controls adverse effects from alloreplete T cells after haploidentical stem cell transplantation. Blood. 2015;125(26):4103-13.

[152] Administration of Donor T Cells With the Caspase-9 Suicide Gene (DOTTI) [Internet]. 2020. Available from: https://clinicaltrials.gov/ct2/show/NCT01494103.

[153] Cam L, Boucquey A, Coulomb-L'hermine A, Weber A, Horellou P. Gene transfer of constitutively active caspase-3 induces apoptosis in a human hepatoma cell line. The Journal of Gene Medicine: A cross-disciplinary journal for research on the science of gene transfer and its clinical applications. 2005;7(1):30-8.

[154] Chen X, Godbey W. 638. Expression-Targeted Cancer Gene Therapy Using pran to Drive Caspase 3 Expression. Mol Ther. 2015;23:S254.

[155] Jia L-T, Chen S-Y, Yang A-G. Cancer gene therapy targeting cellular apoptosis machinery. Cancer Treat Rev. 2012;38(7):868-76.

[156] Zhang D-X, Zhao P-T, Xia L, Liu L-L, Liang J, Zhai H-H, et al. Potent inhibition of human gastric cancer by HER2-directed induction of apoptosis with anti-HER2 antibody and caspase-3 fusion protein. Gut. 2010;59(3):292-9.

[157] Ando M, Hoyos V, Yagyu S, Tao W, Ramos CA, Dotti G, et al. Bortezomib sensitizes non-small cell lung cancer to mesenchymal stromal cell-delivered inducible caspase-9-mediated cytotoxicity. Cancer Gene Ther. 2014;21(11):472-82.

[158] Hoyos V, Del Bufalo F, Yagyu S, Ando M, Dotti G, Suzuki M, et al. Mesenchymal stromal cells for linked delivery of oncolytic and apoptotic adenoviruses to non-small-cell lung cancers. Mol Ther. 2015;23(9):1497-506.

[159] Finlay D, Teriete P, Vamos M, Cosford N, Vuori K. Inducing death in tumor cells: roles of the inhibitor of apoptosis proteins [version 1; referees: 3 approved]. F1000Research. 2017;6(587).

[160] Cong H, Xu L, Wu Y, Qu Z, Bian T, Zhang W, et al. Inhibitor of apoptosis protein (IAP) antagonists in anticancer agent discovery: Current status and perspectives. J Med Chem. 2019;62(12):5750-72.

[161] Infante JR, Dees EC, Olszanski AJ, Dhuria SV, Sen S, Cameron S, et al. Phase I dose-escalation study of LCL161, an oral inhibitor of apoptosis proteins inhibitor, in patients with advanced solid tumors. J Clin Oncol. 2014;32(28):3103-10.

[162] Bardia A, Parton M, Kümmel S, Estévez LG, Huang C-S, Cortés J, et al. Paclitaxel with inhibitor of apoptosis antagonist, LCL161, for localized triple-negative breast cancer, prospectively stratified by gene signature in a biomarker-driven neoadjuvant trial. J Clin Oncol. 2018;36(31):3126-33.

[163] Noonan AM, Bunch KP, Chen JQ, Herrmann MA, Lee JM, Kohn EC, et al. Pharmacodynamic markers and clinical results from the phase 2 study of the SMAC mimetic birinapant in women with relapsed platinum-resistant or-refractory epithelial ovarian cancer. Cancer. 2016;122(4):588-97.

[164] Amaravadi RK, Schilder RJ, Martin LP, Levin M, Graham MA, Weng DE, et al. A phase I study of the SMAC-mimetic birinapant in adults with refractory solid tumors or lymphoma. Mol Cancer Ther. 2015;14(11):2569-75.

[165] Mohamed MS, Bishr MK, Almutairi FM, Ali AG. Inhibitors of apoptosis: clinical implications in cancer. Apoptosis. 2017;22(12):1487-509.

[166] Najem S, Langemann D, Appl B, Trochimiuk M, Hundsdoerfer P, Reinshagen K, et al. Smac mimetic LCL161 supports neuroblastoma chemotherapy in a drug class-dependent manner and synergistically interacts with ALK inhibitor TAE684 in cells with ALK mutation F1174L. Oncotarget. 2016;7(45):72634.

[167] Ziegler DS, Keating J, Kesari S, Fast EM, Zawel L, Ramakrishna N, et al. A small-molecule IAP inhibitor overcomes resistance to cytotoxic therapies in malignant gliomas in vitro and in vivo. Neuro Oncol. 2011;13(8):820-9.

Chapter 2

The Program Cell Death (Apoptosis) and the Therapy of Cancer

Hilal Kalkan

Abstract

Apoptosis plays many vital roles in maintaining organ homeostasis and represents type I programmed cell death. Programmed cell death happens when the DNA damage is irremediable and has two important pathways, the intrinsic death pathway also known as the mitochondrial pathway, and the extrinsic programmed cell death pathway. Any defects in the regulation of these crucial pathways have been associated with many disorders, most importantly cancer. Therefore, understanding the molecular basis of apoptosis is essential for the treatment of incurable cancer. To date, several anti-cancer drugs have been developed by targeting anti-apoptotic proteins, which are upregulated in many cancers. Nonetheless, a disease progression often time warranted due to the deregulation of several anti or pro-apoptotic proteins which also contribute to drug resistance. Hence, it is important to understand the maintenance and counteraction of apoptosis and improve successful new pharmacological applications of cell death mechanisms for future therapies. This chapter discusses the mechanism of apoptosis and emerging principles of drug resistance in cancer.

Keywords: Apoptosis, Caspases, Cell death pathways, Drug resistance, Cancer

1. Introduction

Apoptosis is a type of programmed cell death that is a genetically controlled process of self-destruction. Upon appropriate internal and external stimulus, apoptosis is activated to eliminate the unwanted cells. Hence, damaged cells undergo DNA condensation, membrane blebbing, and cell shrinkage which is the result of activated caspases and proteases [1]. Programmed cell death (PCD) or apoptosis plays an important role to maintain tissue homeostasis. The major aspect of the PCD mechanism is the removal of infected, damaged, or mutation carrying cells (that cause cell-cycle defects). Apoptosis or type I programmed cell death is a well-defined regulated mechanism for the elimination of unwanted cells in the later-period of life. Apoptosis happens in a silent-manner, without causing any damaging effect on the neighboring cells, tissues, and so extinguish stimulation of immunological response [2]. Although, unregulated or deficit form of apoptosis can lead to disease conditions such as cancer and autoimmune diseases [3]. Therefore, understanding the physiological control of apoptosis is crucial for the development of novel approaches to target various diseases. In this chapter, we will provide a brief introduction of apoptosis, the program cell death pathways, the role of caspases, the role of PCD in cancer, and anti-cancer therapy.

2. Apoptosis: type 1 program cell death

Apoptosis (will be referred to as PCD) as the name suggest is the targeted killing of cells that occurs as a part of homeostasis, development, and pathogenic processes throughout life. PCD is well-operated in the physiological processes in an orchestrated manner to eliminate any defect in cells or tissues [2, 4]. The three well-recognized apoptotic cell disassembly are (i) cell shrinkage, (ii) membrane blebbing, and (iii) the formation of membrane protrusions and fragmentation [2]. Apoptosis can be triggered via two ways; i) the extrinsic pathway where the cell death receptors such as Fas, TNFαR, and Death receptors (DR3, DR4, and DR5) bind to their ligands ii) the intrinsic pathway when Bcl-2 family pro-apoptotic proteins lead to permeabilization of the mitochondrial outer membrane. Both extrinsic and intrinsic pathways are converged on the activation of the caspase family, which is responsible for the removal of the damaged cell [5, 6]. The stimulation of apoptosis via the extrinsic pathway requires classic ligand-cell-surface receptor interaction. Death-receptors-induced extrinsic cell death is mainly critical for the function of the immune system [7]. The intrinsic apoptotic cell death starts in a cell-autonomous manner. Cellular defects and stresses, such as DNA damage (due to toxic agents or defect s in the cell-cycle process) or endoplasmic reticulum (ER) stress (caused by accumulated unfolded proteins), leads to apoptosis when the damage is further than repair [8].

2.1 The activation of caspases, functions, and regulation

The initiation of apoptosis requires the activation of caspases, which then promote a cascade of events culminating in the death of a cell. Caspases are a family of cysteine proteases, very well conserved proteins playing many vital roles. Caspases can be divided into two categories (i) involved in immunity and (ii) the ones facilitating apoptotic cell death. These enzymes are expressed inactively inside the cell and activated to cleave for specific target substrates [9]. Pro-apoptotic caspases cleave a much broader subset of intracellular proteins to intercede cell death via apoptosis, on contrary, the caspases those involved in immunity can remove the cells via other non-apoptotic mechanisms. Pro-apoptotic caspases are fundamental for the development, and maintenance of tissue homeostasis [10]. Although many different caspases are found in mammals, here attention will be given to those which are very well characterized. During program cell death, the activation of caspases leads to biochemical and structural changes such as condensation of chromatin, mitochondria permeabilization, and membrane blebbing which are characteristic of apoptotic cell bodies. These essential processes display two important functions, firstly, the promotion of cell death and secondly, the elimination of damaged cells. Caspase-dependent program cell death generally ends with cell death [11]. On the contrary, the failure of apoptosis or non-complete process cause many disorders, most importantly, cancer. Nonetheless, cells have many other regulators that positively or negatively maintaining the activation and/or function of caspases [12]. The expression of caspases differs with stimuli type and expression level. While some caspases are expressed following a death trigger, the others are expressed constantly but being in an inactive state [13, 14]. Once the cell senses a death signal and/or being damaged the initiator caspases are being activated and recruited to the plasma membrane or in the cytoplasm. Upon the cleavage of initiator caspases, they activate the destructor caspases. The activation of destructor caspases in turn cleave important cellular substrates during program cell death. The initiator caspases (caspase −8, −9, and − 10) and destructor caspases (caspase −3, −6, and − 7) prefer different cleavage sites [15]. Many of the destructor caspases include a cleavage site that is consistent with the peptide substrate specificity. Even though, caspase −3 and − 7 contain identical peptide substrate preferences they cleave different substrates. Such as Bid, Caspase −6 and − 9, Bid, XIAP, and gelsolin are cleaved by caspase −3, while p23 is cleaved by caspase-7 [16]. The destructor caspases can cleave more than thousands of substrates to enable program cell death. Also, the initiator caspases have many additional important mediators for program cell death. The activation of initiator caspase-8 is through the extrinsic program cell death pathway which is mediated via death receptors [17]. The other initiator caspase-9 is activated by the intrinsic program cell death pathway which is regulated by the apoptosome complex. The active initiator caspases (both −8, and − 9) can cleave downstream destructor caspases (caspase −3, −6, and − 7) [18]. The activation of initiator caspases can be via several cell-death signaling complexes. These different signaling complexes respond to many intra- and extracellular events, which induce the recruitment of initiator caspases for the activation process [19, 20]. It is important to mention that the interactions of caspase-activation platforms and those caspases indicate a death-fold domain in the proteins. These sorts of forms are existing in caspases and adaptor proteins: the caspase-recruitment domain (CARD) and death effector domain (DED). There are also different other death domains, such as the death domain (DD), and pyrin domain (PYR), which are included in some of the activations of caspases but absent on caspases [21]. Indeed, death fold domains generally intervene in interactions between proteins via homotypic interaction. For example, some extracellular ligands are tumor-necrosis factor (TNF) family, TNF-related apoptosis-inducing ligand (TRAIL), and FasL. These specific ligands bind to different TNF receptor (TNFR) family, so, upon their binding stimulating conformational changes. The specific signaling complex that is linked with Fas is DISC (death-inducing signaling complex) and contains FADD (Fas-associated death domain), which is a small adaptor protein. TNFR induces the activation of several cell-death signaling complexes [22]. Complex IIa involves FADD and TRADD, and the other complex IIb involves FADD, and RIPK1, RIPK3 (receptor-interacting serine–threonine kinases). Both complexes can activate caspase-8 and program cell death [23]. Once caspase-9 is activated, the cell death stimuli induce Bcl-2 family proteins Bax or Bak and lead to mitochondrial outer membrane permeabilization (MOMP) [24]. Later, MOMP is freeing cytochrome-c (Cyt-c) and other pro-death factors from the mitochondrial membrane to the cytosol [25]. The electrons are carried by Cyt-c in the respiratory chain of healthy cells, which function as a co-factor of Apaf-1 in the apoptotic cell's cytosol. The Cyt-c and ATP bind to Apaf-1 to form the apoptosome complex, which results in the recruitment and activation of caspase-9 [26]. The active caspase-9 cleaves, and in turn, induces the activation of caspase-3. Importantly, this intrinsic program cell death pathway is related to a death-receptor pathway through Bcl-2 family member, Bid. The cleavage of Bid through death-receptor activated caspase-8 leads to truncated Bid, which in turn causes the activation of Bax/Bak-induced MOMP [27]. Hence, result in the activation of caspase-9. Caspase-2 activation induced by the PIDDosome protein complex involves the Rip-associated protein with a death domain (RAIDD) and p53-induced protein with a death domain (PIDD) [28]. The activation of caspase-1 in turn cleaves pro-IL-1b to cause inflammation [23]. Different inflammasomes form particular responses to different stimuli such as Nod-like receptors (NLRs) including NLRP1, NLRP3, and NLRC4 [29–31]. The NLR inflammasome protein contains an adapter binding domain and adaptor apoptosis-associated speck-like protein (adaptor ASC) containing CARD. Once ASC oligomerizes, it, in turn, leads to the recruitment of caspase-1 and activates caspase-1 [23]. The activation process of caspases has many violent consequences. Thus, the regulation is tightly managed. The cFLIP (cellular FLICE inhibitory protein), which is caspase-8's inactive mimic, binds to signaling complexes that are associated with death receptors to repress caspase-8 [32]. XIAP (X-linked inhibitor of apoptosis) is the only active site inhibitor of caspases and contains three BIR (baculovirus IAP repeat) motifs. The region BIR2 inhibits the activation of destructor caspases, caspase-3 and -7 via two different mechanisms. The first one, by binding the catalytic side of caspase-3 and -7, and the second one is by binding the newly cleaved protein terminus [33]. Also, the BIR3 motif inhibits the caspase-9 and blocks the dimerization of caspase-9. Of note, the regulation of caspases can be post-transcriptionally by phosphorylation. Several studies reported to inhibit the activity of caspase-2, −8, and − 9 [34]. Besides, caspases can play roles in diverse signaling pathways by their constant activation, for example, some of the specific morphological features of apoptotic cells are due to caspase-intervened activation of some actin cytoskeleton modulators such as p21-activated kinase 2 (PAK2), Rho-associated kinase I (ROCKI) and gelsolin [35–37] (**Figure 1**).

2.2 The mitochondrial pathway of apoptosis (the intrinsic pathway)

The intrinsic pathway contains the let off of cytochrome c from the mitochondria to the cytoplasm in response to cellular stress [26]. The cellular stress by the variety of factors such as DNA damage, deprivation of growth factors, defective ER, and several developmental factors activate the intrinsic pathway. In the mitochondrial pathway, the cleavage and activation of caspases begin with caspase-9, which leads to apoptosome machinery activation [19, 20]. The apoptosis protease-activating factor1 (APAF1) is a key molecule in the intrinsic pathway of apoptosis, besides, an important target for its pro-survival member, and scaffolding the assembled apoptosome [38]. In the Intrinsic apoptosis pathway, firstly, cytochrome c is discharged from mitochondria and enters into the cytosol, by ending with its

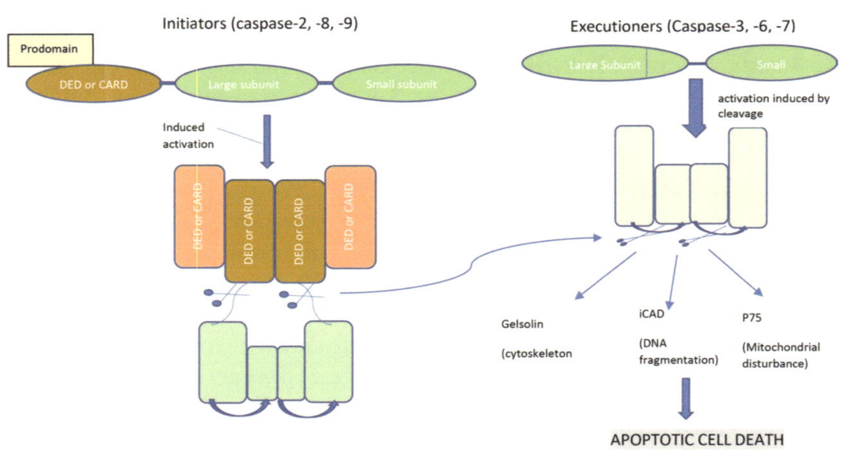

Figure 1.
The caspase family proteases and apoptosis. The initial signal on the cell membrane or from the mitochondria activates the initiator caspases (caspase-2, caspase-8, and caspase-9), which act on execution caspases. The initiator caspases are produced inactively and composed of a pro-domain (containing a CARD or a death effector domain [DED]) and, a large and small subunit. On contrary, executioner caspases (caspase-3, caspase-6, and caspase-7) are activated by cleavage of the initiator caspases. Thus, dependent on initiator signaling. The active caspases including two small and large subunits.

interaction to APAF1 [39, 40]. The cytochrome c, together with APAF1 forms the apoptosome, which leads to inactive caspase-9 monomers into active auto-process [26]. The active caspase-9, in turn, leads to activation of downstream effector caspases. Further, the mitochondria let off pro-apoptotic proteins includes apoptosis-inducing factor (AIF) and endonuclease-G [41]. Cytochrome c is only present in the mitochondrial intermembrane in healthy cells, to interact with APAF1, the apoptotic stimuli trigger the mitochondrial outer membrane (MOM) permeabilization (MOMP) [27]. Hereby, the MOMP conduce to discharge of the entire soluble proteins from mitochondrial intermembrane space to the cytosol. Additionally, rather than cytochrome c, two important pro-apoptotic proteins which are Smac and Omi, were also released meanwhile the above-mentioned procedure. These two proteins are crucial for increasing the potential of apoptosome via antagonizing the caspase inhibitor called X-linked inhibitor of apoptosis (XIAP) [26, 42]. In the lack of Smac and/or Omi, XIAP binding to caspase-9, and inhibits its catalytic activity in which also affects the caspase-3 and caspase-7 [42]. The intrinsic apoptotic pathway is regulated by the Bcl-2 family of proteins [5, 43]. The regulation of MOMP is tightly controlled by the Bcl-2 family proteins, they share BH region homology with their sequence, function, and structure. The Bcl-2 family can be sub-grouped into three categories: the pro-apoptotic proteins, Bax and Bak, play a key role for MOMP regulation; the anti-apoptotic proteins, Bcl-2, Bcl-xL, and Mcl-1, these are the blockers of MOMP; and the BH3-only proteins, Bid, Bim, Bad and Noxa, which are activators of pro-apoptotic proteins and neutralizers for Bcl2 anti-apoptotic proteins. Anti-apoptotic Bcl2 proteins control many other cellular processes such as autophagy and mitochondrial fusion [24, 44]. Bax and Bak are in charge of the loss of mitochondrial outer membrane integrity. Over their activation, they form broad oligomers, and by disturbing MOM enduing their entry into MOM. The disturbing process still unknown, but this lets off the release of entire intermembrane space proteins [45, 46]. Bax and Bak are acting on MOMP involuntarily but importantly at least one of them is needed for re-permeabilization of mitochondria. These proteins are generally being at an inactive state, and gets activated upon upstream stimuli [47–49]. The BH3-only proteins, at least two of them, Bim and active Bid activate Bax and Bak via their transient interaction [25, 50–52]. The antagonization of MOMP accrued by the Bcl2 anti-apoptotic proteins in which they bind and inhibit both Bax and Bak, also BH3-only proteins via interacting with their BH3-only domains [53]. The Bcl2 proteins are regulated both transcriptionally and post-transcriptionally. Especially, via the ubiquitin-proteasome system, the degradation of Mcl-1 involve in apoptosis due to many cellular stresses. Over DNA damage, Mcl1 ubiquitin ligase E3 (MULE) binds to Mcl1 and initiates it catalyzes [54]. Mcl1 is phosphorylated via mitogen-activated protein kinase (MAPK) upon cellular stress conditions [55]. Also, Mcl1 is degraded with growth factors stimuli such as IL3 and the deficiency of phosphoinositide 3-kinase (PI3K)-AKT signaling. This important process eases glycogen synthase kinase 3 (GSK3) from AKT inhibition. Then, GSK3 leads to phosphorylation of Mcl1 and its subsequent degradation [55, 56]. Intrinsic pathway signaling with pro-and anti-apoptotic proteins maintain many signaling processes in the important point of life and death decision. BH3-only proteins are essential upstream sensors of the mitochondrial apoptotic pathway. The regulation of several signaling pathways converge on the BH3-only proteins, and therefore, control and regulate their expression levels and activity [57, 58]. Such as the activation of Bid over cleavage by caspase-8 follow up DR ligation [59, 60]. Hereby, the Bid mediates the cross-regulation of both extrinsic and intrinsic apoptotic pathways. Also, the response to genotoxic stress can be given with Puma and Noxa, which are the direct target of p53 [61–64]. Also, the primary apoptotic factor Bim similarly can be up-regulated via forkhead transcription factor FOXO3A over cytokine deficiency or due to ER stress to unbind the accumulated proteins [65–67]. To this end, the activity of Bad regulated via several kinases' phosphorylation including AKT, which leads to its sequestration through 14–3-3- proteins [68, 69]. Also, when Bad is released, it acts against Bcl2 anti-apoptotic proteins over growth factor deficiency and/or lacking the AKT signaling [68, 70]. It is noteworthy that MOMP is obliged the cell to death even though the caspase activation is hindered which might be the conclusion of the loss of mitochondrial function, which causes the failure of energy [26, 71]. Some cells can survive under these conditions though [26]. Nevertheless, the MOMP activation through its downstream is an important point in the regulation of caspases and might be settled. Finally, mitochondrial apoptosis can be harassed through the phosphorylation and inhibition of caspase-9 by acting downstream of MOMP [72].

2.3 The extrinsic pathway of apoptosis

The extrinsic pathway of apoptosis can be initiated through the binding of ligands to the death receptors such as tumor necrosis factor receptors (TNFRs) and Fas receptor (FasR) on the cell surface [7, 17, 73]. These receptors possess death domains at the cytosolic side, which helps in recruiting adaptor proteins such as FAS associated via death domain (FADD), and also, for some receptors, TNFRSF1A is associated via death domain (TRADD) [74, 75]. The recruitment of these adaptor proteins occurs through the interaction of death domains present in both receptors and adaptors. The death effector domains of these adaptor proteins have different upstream procaspases such as caspase-8 and caspase-10 to form a death-inducing signaling complex (DISC), which then promote activation of these caspases [17, 76]. The downstream executioner caspases such as caspase-3, caspase-6, and caspase-7 activated through their upstream caspases to execute cell death [77]. The main feature of apoptosis is the lack of systemic damage to neighboring cells during its initiation and clearance of the dead cells. This feature of apoptosis is provided by the attentive wrapping of the cell membrane around the fragmented cell by forming apoptotic bodies. The release of ATP (adenosine 5′-triphosphate) and UTP (uridine- 5′-triphosphate) from apoptotic bodies display a signal for macrophages, leads to the migration of macrophages to the apoptotic site [78]. Besides, apoptotic cells also expose the eat me signal most likely phosphatidylserine (PtdSer) on their surface, which initiates phagocytes to engulf the must-remove cells [79].

3. Apoptosis and cancer

Cancer is a multi-complex disease arising from mutation of tumor suppressor or oncogenes, which can be due to impairment of several signaling pathways and associated with apoptosis [80]. The evasion of apoptosis is an important sign of cancer, which mostly results in a very high level of resistance to many chemotherapeutic agents. Therefore, attention has been given to apoptotic proteins and became the most attractive anti-cancer target for effective treatment. The resistance against apoptosis is seen in many types of cancers, and required much higher doses of drugs to overcome the resistance, which increases the risk of "off-target" negative effects [81]. Apoptosis is also a very complex system with its regulation and/or function via several proteins, kinases, and links with several pathways might provide important therapeutic benefit for novel approaches to treat cancer. A better understanding of apoptotic pathways and their alteration in cancer cells had the attention of the development of pro-apoptotic agents. In cancer cells, cell death has been reported for a long time and associated with cancer therapy due to the designation of killing malignant cells by chemo- and/or radiotherapy. Unfortunately, this also leads to the death of normal cells along with cancer cells [80]. Kerr et al. reported that radiation increases the rate of apoptosis in cancer cells. Therefore, apoptosis-associated directly with the development of cancer cells in humans. Most importantly, it can also promote resistance to the therapies which aim to kill cancer cell by targeting apoptosis [2]. In healthy cells, the recognition of damage leads to activation of program cell death, which ending up with their elimination by killing themselves. Such as the activation of tumor suppressor p53 can activate apoptosis inducers including Puma and Noxa, in normal cells [61, 62, 64]. Thus, the un-endowed cells which are not able to eliminate themselves may cause an accumulation of genetic damage. The oncogenes (e.g. Myc) induced apoptosis is a very crucial protective mechanism against cancer development, controlled by a p53-dependent pathway, activated in response to abnormal mitogenic signals. Such as the cells overexpressing the apoptosis inhibitor Bcl2 or have damage in the p53 gene are incapable of this removal process including cancer cells. The loss of function mutations in the p53 tumor suppressor gene causes defection in the apoptotic mechanism and also impacts its tumor-suppressive functions, which involving DNA repairment, cell-cycle arrest, and cell senescence [82, 83]. Also, the incapable cells with an additional mutation can lead to atypical cell proliferation. Such as chromosome translocation which ends up with over-expression of c-Myc, then will lead to quick malignant cell growth. Of note, this explains that the defects in apoptotic mechanism such as over-expression of apoptosis inhibitor Bcl-2 and unregulated cell proliferation due to c-Myc overexpression in the development of lymphoma [82, 84, 85].

Understanding of extrinsic apoptotic pathway on death receptors especially on TRIAL signaling, including TRAIL-1 and -2, shown to lower toxicity compared to TNF or FAS signaling were targeted. The lowering expression levels of TRAIL receptors via recombinant human TRAIL-R1 and -R2 opened up an important therapeutic window. The tumor cells pre-dominantly signal via TRAIL-R1 and -R2, and the expression of these receptors displayed that they do not behave as an indicator of favored receptor signaling [86]. The higher expression levels of these receptors have been found on brain tissue and hepatocytes [87, 88]. However, these findings opened up several windows about the use of TRIAL in the clinic which has been reported notable apoptotic cell death in primary hepatocytes. These findings underlining the importance of selective models to evaluate appropriate drug combinations and their potential on- and/or off-target adverse effects. A study by Von Karstedt et al. in KRAS mutant cancers, known to have an elevated level of TRAIL-R2 expression shown that triggering TRAIL signaling is worsen the development of cancer, and metastasis [89, 90]. Several other studies in different cancers such as pancreatic cancer also shown that TRIAL-R2 signaling interacts with the miRNA processing complex and leads to the inhibition of the maturation of mRNA, called let-7. The result of TRAIL R2-miRNA interaction displayed an up-regulated level of let-7 transcriptional target genes which also prompt cancer development [91]. The sublethal concentrations of TRAIL were shown to induce the activation of caspase-8 dependent apoptotic nucleases, which leads to increased mutations caused by DNA damage in surviving cells [92].

The IAP proteins which are downstream of DRs and Bcl2 family proteins are excessively expressed in cancer cells, therefore associated with cancer development and as well as resistance to therapies. The involvement of cIAPs has also been associated with the regulation of the NF-KB pathway. The lack of cIAPs leads to an increase of inflammatory cytokines including TNF family and IL6, IL8, IL10, and Mcl1, which are effectively involved in systemic inflammation [93]. Nevertheless, these cytokines have also been reported to be increased in cancer progression due to their link with inflammation which causes to include immune cells to the cancer site and prompt the tumor development (**Figure 2**).

3.1 p53 phosphorylation: is it a driver to cell death?

DNA damage comprises different causes that involve cell stress, mutations, toxic agents, or genomic lesions. Upon damage, the confrontation of cells to damage lead to their potential responses to repair the damage or eliminate the harmed cells by apoptosis. Usually, the response of the damaged cell fate differs from strength to the

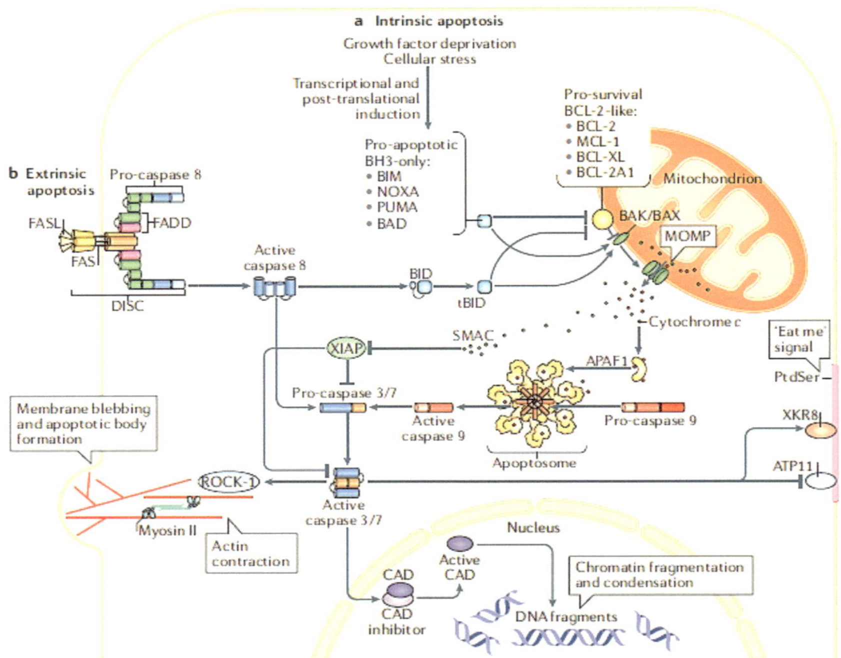

Figure 2.
Molecular mechanisms of intrinsic and extrinsic apoptotic signaling pathways activation. **a.** *the activation of intrinsic apoptotic signaling pathway through cellular stress (ER), and deprivation of growth factor (DNA damage). This alteration promoting the activation of pro-apoptotic members of the BCL-2 family, including BIM, PUMA, BAD, or NOXA. The BH3-only proteins bind and therefore, neutralize BCL-2 pro-survival proteins, involving BCL-2, BCL- XL, MCL-1, and BCL-2- related protein A1. Thus, releasing the important effectors of apoptosis signaling, such as BAK and BAX, which then unite into big complexes and lead to incompliance in the mitochondrial outer membrane. Afterward, this results in the free off cytochrome c and SMAC, which are apoptogenic factors. Some of BH3- only proteins (BIM and PUMA) recently have been reported to bind and promote the activation of MOMP. Also, cytochrome c binding to APAF1 leads to the formation of the apoptosome complex. Caspase-9 is activated in this apoptosome complex, and later proteolytically causes the activation of both caspase- 3 and caspase-7. The caspase inhibitor XIAP activity is blocked by SMAC.* **b.** *Extrinsic apoptotic signaling is stimulated through the activation of DRs, for example, the activation of FAS via FAS ligand (FASL). This cause the pro-caspase 8 recruitment to the death receptor by the adaptor protein FADD, and ending up with the formation of the DISC, which is an important catalyzer for the caspase-8 activation. The activation of caspase 8 promotes cell killing, in two ways: i) via direct activation of caspases 3 and 7, or ii) by indirect activation of the BH3 only protein BID into tBID, thus linking with the intrinsic apoptotic signaling pathway. Effector caspases can be activated in two ways, either by cleavage of hundreds of intracellular proteins to promote the conventional apoptotic morphology and to avoid the efflux of intracellular DAMPs which can result in inflammatory responses. In addition, effector caspases either directly or indirectly can lead to activation of ROCK-1 kinase, which is important for plasma membrane blebbing by actin contraction and caspase-activated DNAse (CAD), which cause inter-nucleosomal DNA cleavage and thus chromatin condensation. These effector caspases also can proteolytically lead to the inactivation of lipid flippases including ATP11, and also can activate the lipid scramblase such as XKR8. Altogether, this results in the exposure of PtdSer on the cell membrane. Therefore, this is called an 'eat me signal' for phagocytic cells and ending up with the engulfment of cells that are undergoing apoptosis [94].*

content of DNA damage, and different signaling pathways are activated. The different cellular decision-making process is guided with specific signaling molecules at the molecular level [95]. The tumor suppressor p53 plays a key role in response to DNA damage and involve in the cellular decision-making process, in turn, found to prompt DNA repair by shutoff the cell cycle to allow time for the repairment and restore genome stability [96]. However, it remains unknown still that how p53 does different cell-fate decisions, activate different target genes, and at a non-transcriptional level have an effect in the cytoplasm of MOMP [97]. The phosphorylation of p53 at Ser15 and Ser20 occurs over immense DNA damage and these are critical sites for stabilization of p53 [98]. Importantly, there is a specific phosphorylation site which is Ser46 and selectively associated with the activity of cell killing caused by genotoxic stress. These crucial different phosphor-sites of p53 might improve the understanding of the function of specific phosphorylation sites. Many studies focused on the phosphorylation of p53 at Ser46 which modulates the target gene transactivation and it has been implicated to potentiate cell death transcriptionally and also, non-transcriptionally [95]. The transactivation of significant pro-apoptotic p53 target genes including BAX, NOXA, p53AIP1, PTEN, and AKT kinase, which are all up-regulated over p53 Ser46 phosphorylation [63, 99–101]. How this different transcriptional effect occurs and which mechanisms are evolved is an open question? The specific role of the p53 Ser46 phosphorylation site in making a cell-fate decision upon cell death leads to the thought of its disturbed regulation might cause cancer cell resistance to therapy. The deregulation of p53 Ser46 kinases in cancer cells could be caused by many factors involving mutations, mis-localization, lower expression level, and disturbed kinase activity. The mutations of p53 at Ser46 have been reported with 1.4–1.6% rates in different human cancer types including breast, colorectal, lung, and stomach. Besides, the down-regulation of several p53 Ser46 kinases are reported such as ATM, DYRK2, HIPK2, and p38α in different tumor types [63, 102–104]. Although, since the regulation of the kinase relies on p53 Ser46 activity, their specific potentials and activities is a challenging task.

3.2 The pro-oncogenic roles of apoptosis

The deregulation of apoptosis is a feature of cancer cells. Apoptosis is a tumor suppressor mechanism in cancer cells. Hence, the dysregulation of apoptosis causing uncontrollable cancer cell-survival [105]. Besides, the tumor-suppressive effects of apoptosis, also, suggested having pro-oncogenic functions. The maintenance of aggressive tumors by apoptosis has been described in the literature. The term tumor aggressiveness defined with increased proliferation, and metastatic features of cancer cells. Such as apoptotic stresses are maybe the reason for aggressive tumor development. Therefore, have pro-oncogenic roles in cancer development and aggressiveness [106, 107]. Importantly, this opens several questions about the beneficial effects of apoptosis in cancer therapy, since strategies aim to trigger apoptosis [108]. Apoptosis-induced proliferation (AIP) is an important compensatory proliferation form, which helps to eliminate damaged cells and maintain tissue homeostasis. Caspase-dependent inflammation and repressed AIP activity might lead to tumor development. One important mediator of AIP is Prostaglandin-E2 (PGE-2), and functioning as a trigger of tumor proliferation in apoptotic cells [109]. The regulation of tumor cells by PGE-2 through caspase-3 dependent manner is an important strategy to treat melanoma. Here, the dying cells trigger the growth of living-cells after cytotoxic therapy [110, 111]. Therefore, targeting PGE-2 to enhance the response to therapy might help to overcome the resistance in cancer cells, when designing cancer therapies [112].

3.3 The oncogenic roles of apoptosis

Apoptotic cells release the "eat-me" signal which is a signal for phagocytosis. These dynamic signals are including fractalkine (FTK), phosphatidylserine (PS), nucleotides ATP, and UTP [113, 114]. These signals are responsible for immune cell activation. Such as the effects of FTK associated with cell migration, and angiogenesis. Also, these signals by apoptotic cells are thought to induce inflammation. Somehow, in the occurrence of normal apoptotic cell death, inflammation is prevented by efferocytosis. The mechanism of efferocytosis is to eliminate the apoptotic cells by immune cells, macrophages. Importantly, it also plays a key role in cancer cells by creating an immunosuppressive effect to organize several signalization events between tumor cells [115]. For example, a study by Ford and colloquies, in 2015, showed that apoptotic cancer cells induce tumor growth and angiogenesis in B-cell lymphoma and also, malignant melanoma [116]. In conclusion, the effects of apoptotic cells in their microenvironment is changed upon their features. The studies that suggest apoptotic cell involvement might play important role in signaling events due to their environment and features. Of note, it remains unknown how they do these roles, therefore, more studies are needed to clarify this point.

4. The therapeutic approaches to cancer

The question what happens when apoptotic machinery is impaired? The answer is impairment of apoptosis results in uncontrolled cell proliferation, tumor development and progression, and resistance to anticancer therapeutics [117, 118]. The deregulation of apoptosis is not only causing the development and progression of the tumor but also lead to resistance to therapeutic approaches [80]. Cancer cells can promote drug resistance to several treatments, which target several different molecular signaling. So many cancers present an internal resistance to chemotherapy, without previous exposure to anti-cancer agents. Therefore, the beginning response to cancer treatment is weak. Furthermore, several tumors might also gain drug resistance because of chemotherapy [119–122]. Among many intrinsic and gained drug resistance are increased drug leakage, defective inflow transporters, changes in the function of drugs and their targets, drug inactivation metabolically, drug compartmentalization, increased activity of DNA repairment [120, 123–126]. To improve therapeutically new approaches, the understanding of these signaling pathways is required to progress cancer therapy and more importantly get over the resistance. Besides, targeting the apoptotic signaling events that are involved in resistance may improve a valid strategy to increase the cell sensitivity of cancer cells to apoptosis and overcome the failure of therapeutic difficulties.

4.1 The role of Bcl2 family proteins in cancer and drug resistance

The Bcl-2 family proteins contains at least one Bcl-2 homology (BH) domain [127–129]. The different BH-domains exist, but especially BH-3 is crucial to form several BCL-2 protein complexes that govern the result between survival and death [130–133]. These interactions between BH3-BH3 are also important to antagonize pro-survival complexes. The increased level of anti-apoptotic Bcl-2 family proteins is associated with carcinogenesis. Such as the transgenic mice model that overexpress Bcl-2 shown to develop tumors spontaneously [134]. However, the high expression of Bcl-2 was also reported to cause a shift in the balance of cancer cells through survival and also lead to resistance to chemotherapy [135]. Overexpression of Bcl-xL, another member of the Bcl2-family is also associated with colorectal cancer [136]. This unbalanced expression of Bcl-2 family proteins confers failure of apoptosis and resistance to cancer therapies [137]. Especially, overexpression of anti-apoptotic proteins such as Bcl-2 and Bcl-xL is correlated with several cancers including ovarian and breast cancer [138, 139]. On the other side, an increased ratio of Bcl-2/Bax is associated with chronic lymphocytic leukemia (CLL) [140].

4.2 The role of BH3-mimetics as a potential therapeutic option for cancer

Bcl-2 family proteins control the intrinsic apoptosis pathway toward protein–protein interactions, which including all BH-3 domains [141, 142]. Currently, the therapeutic approaches mostly focused on mimicking and target the BH-3 only proteins. BH-3 mimetics are small molecule compounds and bind to anti-apoptotic Bcl-2 proteins. Therefore, inhibit their activity and promotes apoptosis [133, 143]. Of note, BH-3 mimetics can start apoptosis by leading to activation of Bax/Bak or by inhibiting anti-apoptotic Bcl-2 family proteins [142]. Between a variety of BH-3 mimetics that have been generated, ABT-737 and ABT-263 (Navitoclax) showed the most promising antitumor efficacy by engaging anti-apoptotic Bcl-2 family proteins. Both drugs are shown to engage to Bcl-2, Bcl-Xl, and Bcl-w. Somehow, the use of Navitoclax showed to cause thrombocytopenia by inhibition of Bcl-xl, as a major adverse effect [144, 145]. These side effects lead to exclusion of Navitoclax as an effective chemotherapeutic drug and required new findings more selective for Bcl-2 inhibitors, so can induce cell death without causing critical side effects. Indeed, ABT-199 (Navitoclax) was subsequently developed to replace Navitoclax, as a highly selective Bcl-2 inhibitor. Venetoclax binds with high affinity to Bcl-2 and a lesser extent to other Bcl-2 family proteins including Bcl-xl and Bcl-w. Also, Venetoclax lacks off-target effects on BCL-XL [146]. Altogether, the pre-clinical experience with venetoclax and together with combining BCR signaling inhibitors were shown to be very promising with significant efficacy in the patient population Chronic Lymphocytic Leukemia (CLL) and Diffuse Large B-cell Lymphoma (DLBCL). The study by Binu Sasi et al., in 2019, showed that fostamatinib and ibrutinib can sensitize DLBCL cell lines to venetoclax, which marks the importance of combination approaches. In conclusion, the findings provided a rational explanation for testing the use of combination therapy with venetoclax and selective BCR inhibitors in clinical use in DLBCL [147]. The progress of research is underlying the continuous role of Bcl-2 family inhibitors as promising and effective anticancer therapeutic agent. However, another small molecule BDA-366 (BH4 domain inhibitor) is also recently described as a promising approach in anticancer treatment. The potent effects of BDA-366 have been experimentally validated in both in vitro and in vivo, by showing significant anticancer activity [148]. Lastly, a very innovative strategy again comes from Bcl-2, the Bcl-2 gene is known to be rich with GC sequence which is placed in the promoter region, and has the potential to form G-quadruplex structures. Thus, is playing a role as a transcriptional repressor element and these G-quadruplex-specific ligands are capable of regulating Bcl-2 transcriptionally [149, 150].

4.3 Up-regulation of Bcl-2 (pro-apoptotic) family proteins to overcome resistance in cancer cells

The up-regulated expression level of anti-apoptotic Bcl-2 proteins might be possible be neutralized by increasing the expression of pro-apoptotic Bcl-2 proteins. Puma is an important BH-3 protein, having a key role in mediating pro-apoptotic p53 function and help to release p53 from inactive forms with Bcl-Xl. Hereby, freed cytosolic p53 can activate Bax and Bak, which results in stimulation of apoptotic signaling through MOMP [44, 151]. p53 increases the expression of pro-apoptotic protein level, called ARTS, and counteracts Bcl-xl to induce apoptosis [152]. Hence, the inhibition of p53 antagonists such as Mdm-2 can cause an increased level of p53, induce pro-apoptotic protein expressions. Additionally, reduction of pro-apoptotic BH-3 only protein levels can also lead to drug resistance. This effect might be due to its interaction with BCL-2 anti-apoptotic proteins in cancer cells [153].

4.4 Therapeutic approach to target death receptor

There are different reasons and ways cause the failure of apoptosis or leading to resistance. Including altered death receptor signaling, an unbalanced level of pro-and anti-apoptotic proteins lowered caspase role and disrupted p53 function. Impairment of the death receptor signaling pathway is associated with many sorts of tumors, showing the importance of the Fas–FasL system and/or the abnormal level of cytosolic components of this pathway such as FADD can improve the tumor conversion [154, 155]. Transcriptionally silencing of Fas is familiar with oncogenic events in the epithelial conversion, the mutation in this gene linked with B-cell germinal center-derived lymphomas [156]. However, the lack of expression level of FADD is a common sign in acute myelogenous leukemia (AML) which leads to chemotherapy resistance [154, 155, 157]. Furthermore, the low and/or lack of caspase-8 expression have been reported in numerous other cancers including neuroblastoma and small cell lung cancer [158–160]. However, the induction of apoptosis by triggering the death receptor pathway might be an important promising approach to overcome the resistance to therapeutic agents. An important therapeutic approach that confers the TRAIL ligand which induces apoptosis in many different cancer cell lines [161]. TRAIL-induced apoptosis, an important mechanistic strategy that has a harmful effect against cancer cells and leads to several different approaches to develop agonistic-featured antibodies against TRAIL death receptors and/or several soluble recombinant derivatives of TRAIL [162, 163]. On the other side, increasing the sensitivity of malignant tumors to TRAIL-induced apoptosis is crucial to design new promising molecules to target and activate caspase-8 [164]. Importantly, the validation of in silico screening some small molecules which promote caspase-8 activation have been reported. In many different cancer cell lines including leukemic cells, the CaspPro molecule is shown to be very promising, and shown to induce not only caspase-8 activation but also caspase 3 activation, and leading to apoptosis in the presence of TRAIL [163–165].

Among other therapeutic approaches, another important strategy is targeting IAPs. Through the use of antisense oligonucleotides and small-molecule inhibitors very potential anticancer therapeutic approach has been developed. In ovarian cancer, the supplementation of antisense agents targeting XIAP through an adenoviral vector was displayed to reduce the expression level of XIAP and underlined its remarkable effect to induce apoptosis [166]. Besides, in lung cancers also its effect has been proven by showing that it increases cell sensitivity to radiation treatment [167]. In this regard, by inhibition of XIAP through the use of specifically designed oligomers also displayed to promote the activation of caspase-3, increasing the apoptotic effect of TRIAL in different cancers such as prostate cancer [168]. Furthermore, the SMAC mimetics are also shown to trigger apoptosis and/or increase the sensitivity of cancer cells to different therapies. These molecules induce caspase activation through the neutralization of XIAP-mediated caspase inhibition [169]. The successful therapeutic approach relies on the capability of the tool to trigger apoptosis mainly by two approaches, i) targeting overexpressed anti-apoptotic Bcl-2 family proteins, or ii) triggering the activation of pro-apoptotic molecules.

5. Conclusion remark

In conclusion, the main hallmark of cancer is the intrinsic or requisite resistance to apoptosis, and the evasion of apoptotic cell death can be part of cellular response to stressful stimuli. Based on our understanding of how the dysregulation of the apoptosis pathway leads to cancer, approaches to target anti-apoptosis Bcl-2 family proteins have proven to be effective in the treatment of cancers, especially lymphatic cancers. However, cancer cells often outsmart these therapeutic interventions by modulating the expression of other molecules of apoptosis. Novel combination approaches are warranted to overcome this drug-induced resistance. More importantly, the new perspective and pieces of evidence are needed for powerful approaches to overcome drug resistance in cancer cells in clinical use. Therefore, the current knowledge needs a further understanding of how different cellular stresses trigger anti-apoptotic mechanisms, and how these effects lead to apoptotic resistance in cancer cells.

Author details

Hilal Kalkan[1,2]

1 Università degli Studi della Campania Luigi Vanvitelli, Naples, Italy

2 Istituto di Chimica Biomolecolare (ICB-CNR), Naples, Italy

*Address all correspondence to: h.kalkan@icb.cnr.it

References

[1] Kroemer G, Galluzzi L, Vandenabeele P, Abrams J, Alnemri ES, Baehrecke EH, Blagosklonny MV, El-Deiry WS, Golstein P, Green DR *et al.* 2009 Classification of cell death: recommendations of the Nomenclature Committee on Cell Death 2009. Cell Death and Differentiation **16** 3-11. (doi:10.1038/cdd.2008.150)

[2] Kerr JF, Wyllie AH & Currie AR 1972 Apoptosis: a basic biological phenomenon with wide-ranging implications in tissue kinetics. British Journal of Cancer **26** 239-257. (doi:10.1038/bjc.1972.33)

[3] Testa U & Riccioni R 2007 Deregulation of apoptosis in acute myeloid leukemia. Haematologica **92** 81-94. (doi:10.3324/haematol.10279)

[4] Wyllie AH, Kerr JF & Currie AR 1980 Cell death: the significance of apoptosis. International Review of Cytology **68** 251-306. (doi:10.1016/s0074-7696(08)62312-8)

[5] Elmore S 2007 Apoptosis: a review of programmed cell death. Toxicologic Pathology **35** 495-516. (doi:10.1080/01926230701320337)

[6] Sayers TJ 2011 Targeting the extrinsic apoptosis signaling pathway for cancer therapy. Cancer Immunology, Immunotherapy **60** 1173-1180. (doi:10.1007/s00262-011-1008-4)

[7] Locksley RM, Killeen N & Lenardo MJ 2001 The TNF and TNF Receptor Superfamilies: Integrating Mammalian Biology. Cell **104** 487-501. (doi:10.1016/S0092-8674(01)00237-9)

[8] Iurlaro R & Muñoz-Pinedo C 2016 Cell death induced by endoplasmic reticulum stress. The FEBS Journal **283** 2640-2652. (doi:https://doi.org/10.1111/febs.13598)

[9] Teng X & Hardwick JM 2015 Apoptosis: The extrinsic pathway. *Molecular Oncology: Causes of Cancer and Targets for Treatment* 353-366. (doi:10.1017/CBO9781139046947.031)

[10] Puthalakath H & Strasser A 2002 Keeping killers on a tight leash: transcriptional and post-translational control of the pro-apoptotic activity of BH3-only proteins. Cell Death and Differentiation **9** 505-512. (doi:10.1038/sj.cdd.4400998)

[11] Green DR & Llambi F 2015 Cell Death Signaling. Cold Spring Harbor Perspectives in Biology **7**. (doi:10.1101/cshperspect.a006080)

[12] Fulda S 2009a Exploiting apoptosis pathways for the treatment of pediatric cancers. Pediatric Blood & Cancer **53** 533-536. (doi:https://doi.org/10.1002/pbc.21922)

[13] Creagh EM, Conroy H & Martin SJ 2003 Caspase-activation pathways in apoptosis and immunity. Immunological Reviews **193** 10-21. (doi:10.1034/j.1600-065x.2003.00048.x)

[14] Nicholson DW 1999 Caspase structure, proteolytic substrates, and function during apoptotic cell death. Cell Death and Differentiation **6** 1028-1042. (doi:10.1038/sj.cdd.4400598)

[15] Stennicke HR & Salvesen GS 1998 Properties of the caspases. Biochimica et Biophysica Acta (BBA) - Protein Structure and Molecular Enzymology **1387** 17-31. (doi:10.1016/S0167-4838(98)00133-2)

[16] Walsh JG, Cullen SP, Sheridan C, Luthi AU, Gerner C & Martin SJ 2008 Executioner caspase-3 and caspase-7 are functionally distinct proteases. *Proceedings of the National Academy of Sciences* **105** 12815-12819. (doi:10.1073/pnas.0707715105)

[17] Dickens LS, Powley IR, Hughes MA & MacFarlane M 2012 The 'complexities' of life and death: Death receptor signalling platforms. Experimental Cell Research **318** 1269-1277. (doi:10.1016/j.yexcr.2012.04.005)

[18] Crawford ED & Wells JA 2011 Caspase substrates and cellular remodeling. Annual Review of Biochemistry **80** 1055-1087. (doi:10.1146/annurev-biochem-061809-121639)

[19] Bratton SB & Salvesen GS 2010 Regulation of the Apaf-1-caspase-9 apoptosome. Journal of Cell Science **123** 3209-3214. (doi:10.1242/jcs.073643)

[20] Bratton SB, Walker G, Srinivasula SM, Sun XM, Butterworth M, Alnemri ES & Cohen GM 2001 Recruitment, activation and retention of caspases-9 and -3 by Apaf-1 apoptosome and associated XIAP complexes. The EMBO Journal **20** 998-1009. (doi:10.1093/emboj/20.5.998)

[21] Malladi S, Challa-Malladi M, Fearnhead HO & Bratton SB 2009 The Apaf-1•procaspase-9 apoptosome complex functions as a proteolytic-based molecular timer. The EMBO Journal **28** 1916-1925. (doi:10.1038/emboj.2009.152)

[22] Bodmer JL, Holler N, Reynard S, Vinciguerra P, Schneider P, Juo P, Blenis J & Tschopp J 2000 TRAIL receptor-2 signals apoptosis through FADD and caspase-8. Nature Cell Biology **2** 241-243. (doi:10.1038/35008667)

[23] Berghe TV, Linkermann A, Jouan-Lanhouet S, Walczak H & Vandenabeele P 2014 Regulated necrosis: the expanding network of non-apoptotic cell death pathways. Nature Reviews Molecular Cell Biology **15** 135-147. (doi:10.1038/nrm3737)

[24] Cosentino K & García-Sáez AJ 2018 MIM through MOM: the awakening of Bax and Bak pores. *The EMBO Journal* **37** e100340. (doi:10.15252/embj.2018100340)

[25] Kuwana T, Mackey MR, Perkins G, Ellisman MH, Latterich M, Schneiter R, Green DR & Newmeyer DD 2002 Bid, Bax, and Lipids Cooperate to Form Supramolecular Openings in the Outer Mitochondrial Membrane. Cell **111** 331-342. (doi:10.1016/S0092-8674(02)01036-X)

[26] Tait SWG & Green DR 2010 Mitochondria and cell death: outer membrane permeabilization and beyond. Nature Reviews. Molecular Cell Biology **11** 621-632. (doi:10.1038/nrm2952)

[27] Suzuki M, Youle RJ & Tjandra N 2000 Structure of Bax: Coregulation of Dimer Formation and Intracellular Localization. Cell **103** 645-654. (doi:10.1016/S0092-8674(00)00167-7)

[28] Park HH & Wu H 2006 Crystal structure of RAIDD death domain implicates potential mechanism of PIDDosome assembly. Journal of Molecular Biology **357** 358-364. (doi:10.1016/j.jmb.2005.12.082)

[29] Kanneganti T-D, Lamkanfi M & Núñez G 2007 Intracellular NOD-like Receptors in Host Defense and Disease. Immunity **27** 549-559. (doi:10.1016/j.immuni.2007.10.002)

[30] Mariathasan S, Newton K, Monack DM, Vucic D, French DM, Lee WP, Roose-Girma M, Erickson S & Dixit VM 2004 Differential activation of the inflammasome by caspase-1 adaptors ASC and Ipaf. Nature **430** 213-218. (doi:10.1038/nature02664)

[31] Martinon F 2008 Detection of immune danger signals by NALP3. Journal of Leukocyte Biology **83** 507-511. (doi:https://doi.org/10.1189/jlb.0607362)

[32] Budd RC, Yeh W-C & Tschopp J 2006 cFLIP regulation of lymphocyte activation and development. *Nature Reviews*. Immunology **6** 196-204. (doi:10.1038/nri1787)

[33] Riedl SJ, Renatus M, Schwarzenbacher R, Zhou Q, Sun C, Fesik SW, Liddington RC & Salvesen GS 2001 Structural Basis for the Inhibition of Caspase-3 by XIAP. Cell **104** 791-800. (doi:10.1016/S0092-8674(01)00274-4)

[34] Parrish AB, Freel CD & Kornbluth S 2013 Cellular Mechanisms Controlling Caspase Activation and Function. Cold Spring Harbor Perspectives in Biology **5** a008672. (doi:10.1101/cshperspect.a008672)

[35] Coleman ML, Sahai EA, Yeo M, Bosch M, Dewar A & Olson MF 2001 Membrane blebbing during apoptosis results from caspase-mediated activation of ROCK I. Nature Cell Biology **3** 339-345. (doi:10.1038/35070009)

[36] Rudel T & Bokoch GM 1997 Membrane and morphological changes in apoptotic cells regulated by caspase-mediated activation of PAK2. Science (New York, N.Y.) **276** 1571-1574. (doi:10.1126/science.276.5318.1571)

[37] Sebbagh M, Renvoizé C, Hamelin J, Riché N, Bertoglio J & Bréard J 2001 Caspase-3-mediated cleavage of ROCK I induces MLC phosphorylation and apoptotic membrane blebbing. Nature Cell Biology **3** 346-352. (doi:10.1038/35070019)

[38] Zhou P, Qian L, Kozopas KM & Craig RW 1997 Mcl-1, a Bcl-2 family member, delays the death of hematopoietic cells under a variety of apoptosis-inducing conditions. Blood **89** 630-643.

[39] Yu X, Acehan D, Ménétret J-F, Booth CR, Ludtke SJ, Riedl SJ, Shi Y, Wang X & Akey CW 2005 A Structure of the Human Apoptosome at 12.8 Å Resolution Provides Insights into This Cell Death Platform. Structure **13** 1725-1735. (doi:10.1016/j.str.2005.09.006)

[40] Yuan S, Yu X, Topf M, Ludtke SJ, Wang X & Akey CW 2010 Structure of an Apoptosome-Procaspase-9 CARD Complex. Structure **18** 571-583. (doi:10.1016/j.str.2010.04.001)

[41] Toné S, Sugimoto K, Tanda K, Suda T, Uehira K, Kanouchi H, Samejima K, Minatogawa Y & Earnshaw WC 2007 Three distinct stages of apoptotic nuclear condensation revealed by time-lapse imaging, biochemical and electron microscopy analysis of cell-free apoptosis. Experimental Cell Research **313** 3635-3644. (doi:10.1016/j.yexcr.2007.06.018)

[42] Eckelman BP, Salvesen GS & Scott FL 2006 Human inhibitor of apoptosis proteins: why XIAP is the black sheep of the family. EMBO Reports **7** 988-994. (doi:10.1038/sj.embor.7400795)

[43] Edlich F 2018 BCL-2 proteins and apoptosis: Recent insights and unknowns. Biochemical and Biophysical Research Communications **500** 26-34. (doi:10.1016/j.bbrc.2017.06.190)

[44] Chipuk JE, Fisher JC, Dillon CP, Kriwacki RW, Kuwana T & Green DR 2008 Mechanism of apoptosis induction by inhibition of the anti-apoptotic BCL-2 proteins. *Proceedings of the National Academy of Sciences* **105** 20327-20332. (doi:10.1073/pnas.0808036105)

[45] Goldstein JC, Waterhouse NJ, Juin P, Evan GI & Green DR 2000 The coordinate release of cytochrome c during apoptosis is rapid, complete and kinetically invariant. Nature Cell Biology **2** 156-162. (doi:10.1038/35004029)

[46] Muñoz-Pinedo C, Guío-Carrión A, Goldstein JC, Fitzgerald P, Newmeyer DD & Green DR 2006 Different mitochondrial intermembrane space proteins are released during apoptosis in a manner that is coordinately initiated

but can vary in duration. *Proceedings of the National Academy of Sciences* **103** 11573-11578. (doi:10.1073/pnas.0603007103)

[47] Dewson G, Kratina T, Sim HW, Puthalakath H, Adams JM, Colman PM & Kluck RM 2008 To trigger apoptosis, Bak exposes its BH3 domain and homodimerizes via BH3:groove interactions. Molecular Cell **30** 369-380. (doi:10.1016/j.molcel.2008.04.005)

[48] Dewson G, Kratina T, Czabotar P, Day CL, Adams JM & Kluck RM 2009 Bak activation for apoptosis involves oligomerization of dimers via their alpha6 helices. Molecular Cell **36** 696-703. (doi:10.1016/j.molcel.2009.11.008)

[49] Eskes R, Desagher S, Antonsson B & Martinou JC 2000 Bid induces the oligomerization and insertion of Bax into the outer mitochondrial membrane. Molecular and Cellular Biology **20** 929-935. (doi:10.1128/mcb.20.3.929-935.2000)

[50] Kuwana T, Bouchier-Hayes L, Chipuk JE, Bonzon C, Sullivan BA, Green DR & Newmeyer DD 2005 BH3 Domains of BH3-Only Proteins Differentially Regulate Bax-Mediated Mitochondrial Membrane Permeabilization Both Directly and Indirectly. Molecular Cell **17** 525-535. (doi:10.1016/j.molcel.2005.02.003)

[51] Letai A, Bassik MC, Walensky LD, Sorcinelli MD, Weiler S & Korsmeyer SJ 2002 Distinct BH3 domains either sensitize or activate mitochondrial apoptosis, serving as prototype cancer therapeutics. Cancer Cell **2** 183-192. (doi:10.1016/S1535-6108(02)00127-7)

[52] Wei T, Ni Y, Hou J, Chen C, Zhao B & Xin W 2000 Hydrogen peroxide-induced oxidative damage and apoptosis in cerebellar granule cells: protection by Ginkgo biloba extract. Pharmacological Research **41** 427-433. (doi:10.1006/phrs.1999.0604)

[53] Llambi F, Moldoveanu T, Tait SWG, Bouchier-Hayes L, Temirov J, McCormick LL, Dillon CP & Green DR 2011 A unified model of mammalian BCL-2 protein family interactions at the mitochondria. Molecular Cell **44** 517-531. (doi:10.1016/j.molcel.2011.10.001)

[54] Warr MR, Acoca S, Liu Z, Germain M, Watson M, Blanchette M, Wing SS & Shore GC 2005 BH3-ligand regulates access of MCL-1 to its E3 ligase. FEBS Letters **579** 5603-5608. (doi:10.1016/j.febslet.2005.09.028)

[55] Wertz IE, Kusam S, Lam C, Okamoto T, Sandoval W, Anderson DJ, Helgason E, Ernst JA, Eby M, Liu J *et al.* 2011 Sensitivity to antitubulin chemotherapeutics is regulated by MCL1 and FBW7. Nature **471** 110-114. (doi:10.1038/nature09779)

[56] Ulrich Maurer, Ce´line Charvet, Allan S. Wagman, Emmanuel Dejardin, and Douglas R. Green. Glycogen Synthase Kinase-3 Regulates Mitochondrial Outer Membrane Permeabilization and Apoptosis by Destabilization of MCL-1. (doi:10.1016/j.molcel.2006.02.009)

[57] Huang T, Song C, Zheng L, Xia L, Li Y & Zhou Y 2019a The roles of extracellular vesicles in gastric cancer development, microenvironment, anti-cancer drug resistance, and therapy. Molecular Cancer **18** 62. (doi:10.1186/s12943-019-0967-5)

[58] Singh R, Letai A & Sarosiek K 2019 Regulation of apoptosis in health and disease: the balancing act of BCL-2 family proteins. Nature Reviews Molecular Cell Biology **20** 175-193. (doi:10.1038/s41580-018-0089-8)

[59] Li J, Simpson L, Takahashi M, Miliaresis C, Myers MP, Tonks N &

Parsons R 1998 The PTEN/MMAC1 Tumor Suppressor Induces Cell Death That Is Rescued by the AKT/Protein Kinase B Oncogene. Cancer Research **58** 5667-5672.

[60] Luo X, Budihardjo I, Zou H, Slaughter C & Wang X 1998 Bid, a Bcl2 interacting protein, mediates cytochrome c release from mitochondria in response to activation of cell surface death receptors. Cell **94** 481-490. (doi:10.1016/s0092-8674(00)81589-5)

[61] Jeffers JR, Parganas E, Lee Y, Yang C, Wang J, Brennan J, MacLean KH, Han J, Chittenden T, Ihle JN *et al.* 2003 Puma is an essential mediator of p53-dependent and -independent apoptotic pathways. Cancer Cell **4** 321-328. (doi:10.1016/s1535-6108(03)00244-7)

[62] Nakano K & Vousden KH 2001 PUMA, a novel proapoptotic gene, is induced by p53. Molecular Cell **7** 683-694. (doi:10.1016/s1097-2765(01)00214-3)

[63] Oda K, Arakawa H, Tanaka T, Matsuda K, Tanikawa C, Mori T, Nishimori H, Tamai K, Tokino T, Nakamura Y *et al.* 2000 p53AIP1, a potential mediator of p53-dependent apoptosis, and its regulation by Ser-46-phosphorylated p53. *Cell* **102** 849-862. (doi:10.1016/s0092-8674(00)00073-8)

[64] Villunger A, Michalak EM, Coultas L, Müllauer F, Böck G, Ausserlechner MJ, Adams JM & Strasser A 2003 p53- and Drug-Induced Apoptotic Responses Mediated by BH3-Only Proteins Puma and Noxa. Science **302** 1036-1038. (doi:10.1126/science.1090072)

[65] Dijkers PF, Medema RH, Lammers JW, Koenderman L & Coffer PJ 2000 Expression of the pro-apoptotic Bcl-2 family member Bim is regulated by the forkhead transcription factor FKHR-L1. Current Biology: CB **10** 1201-1204. (doi:10.1016/s0960-9822(00)00728-4)

[66] Puthalakath H, O'Reilly LA, Gunn P, Lee L, Kelly PN, Huntington ND, Hughes PD, Michalak EM, McKimm-Breschkin J, Motoyama N *et al.* 2007 ER Stress Triggers Apoptosis by Activating BH3-Only Protein Bim. Cell **129** 1337-1349. (doi:10.1016/j.cell.2007.04.027)

[67] Urbich C, Knau A, Fichtlscherer S, Walter DH, Brühl T, Potente M, Hofmann WK, de Vos S, Zeiher AM & Dimmeler S 2005 FOXO-dependent expression of the proapoptotic protein Bim: pivotal role for apoptosis signaling in endothelial progenitor cells. FASEB Journal: Official Publication of the Federation of American Societies for Experimental Biology **19** 974-976. (doi:10.1096/fj.04-2727fje)

[68] Datta SR, Dudek H, Tao X, Masters S, Fu H, Gotoh Y & Greenberg ME 1997 Akt Phosphorylation of BAD Couples Survival Signals to the Cell-Intrinsic Death Machinery. Cell **91** 231-241. (doi:10.1016/S0092-8674(00)80405-5)

[69] del Peso L, González-García M, Page C, Herrera R & Nuñez G 1997 Interleukin-3-induced phosphorylation of BAD through the protein kinase Akt. *Science (New York, N.Y.)* **278** 687-689. (doi:10.1126/science.278.5338.687)

[70] Zha J, Harada H, Yang E, Jockel J & Korsmeyer SJ 1996 Serine phosphorylation of death agonist BAD in response to survival factor results in binding to 14-3-3 not BCL-X(L). Cell **87** 619-628. (doi:10.1016/s0092-8674(00)81382-3)

[71] Lartigue L, Kushnareva Y, Seong Y, Lin H, Faustin B & Newmeyer DD 2009 Caspase-independent Mitochondrial Cell Death Results from Loss of Respiration, Not Cytotoxic Protein Release. Molecular Biology of the Cell

20 4871-4884. (doi:10.1091/mbc.e09-07-0649)

[72] Allan LA & Clarke PR 2009 Apoptosis and autophagy: Regulation of caspase-9 by phosphorylation. The FEBS Journal **276** 6063-6073. (doi:https://doi.org/10.1111/j.1742-4658.2009.07330.x)

[73] Nagata S 1997 Apoptosis by death factor. Cell **88** 355-365. (doi:10.1016/s0092-8674(00)81874-7)

[74] Kischkel FC, Hellbardt S, Behrmann I, Germer M, Pawlita M, Krammer PH & Peter ME 1995 Cytotoxicity-dependent APO-1 (Fas/CD95)-associated proteins form a death-inducing signaling complex (DISC) with the receptor. The EMBO Journal **14** 5579-5588. (doi:10.1002/j.1460-2075.1995.tb00245.x)

[75] Newton K & Dixit VM 2012 Signaling in innate immunity and inflammation. Cold Spring Harbor Perspectives in Biology **4**. (doi:10.1101/cshperspect.a006049)

[76] Boatright KM & Salvesen GS 2003 Mechanisms of caspase activation. Current Opinion in Cell Biology **15** 725-731. (doi:10.1016/j.ceb.2003.10.009)

[77] Billen LP, Kokoski CL, Lovell JF, Leber B & Andrews DW 2008 Bcl-XL Inhibits Membrane Permeabilization by Competing with Bax. PLOS Biology **6** e147. (doi:10.1371/journal.pbio.0060147)

[78] Medina CB & Ravichandran KS 2016 Do not let death do us part: 'find-me' signals in communication between dying cells and the phagocytes. Cell Death & Differentiation **23** 979-989. (doi:10.1038/cdd.2016.13)

[79] Nagata S 2018 Apoptosis and Clearance of Apoptotic Cells. Annual Review of Immunology **36** 489-517. (doi:10.1146/annurev-immunol-042617-053010)

[80] Hanahan D & Weinberg RA 2011 Hallmarks of cancer: the next generation. Cell **144** 646-674. (doi:10.1016/j.cell.2011.02.013)

[81] Fox JL & MacFarlane M 2016 Targeting cell death signalling in cancer: minimising "Collateral damage." *British Journal of Cancer* **115** 5-11. (doi:10.1038/bjc.2016.111)

[82] Janic A, Valente LJ, Wakefield MJ, Di Stefano L, Milla L, Wilcox S, Yang H, Tai L, Vandenberg CJ, Kueh AJ *et al.* 2018 DNA repair processes are critical mediators of p53-dependent tumor suppression. Nature Medicine **24** 947-953. (doi:10.1038/s41591-018-0043-5)

[83] Vousden KH & Lane DP 2007 p53 in health and disease. Nature Reviews Molecular Cell Biology **8** 275-283. (doi:10.1038/nrm2147)

[84] Strasser A, Harris AW, Bath ML & Cory S 1990 Novel primitive lymphoid tumours induced in transgenic mice by cooperation between myc and bcl-2. Nature **348** 331-333. (doi:10.1038/348331a0)

[85] Vaux DL, Cory S & Adams JM 1988 Bcl-2 gene promotes haemopoietic cell survival and cooperates with c-myc to immortalize pre-B cells. Nature **335** 440-442. (doi:10.1038/335440a0)

[86] MacFarlane M, Kohlhaas SL, Sutcliffe MJ, Dyer MJS & Cohen GM 2005 TRAIL Receptor-Selective Mutants Signal to Apoptosis via TRAIL-R1 in Primary Lymphoid Malignancies. Cancer Research **65** 11265-11270. (doi:10.1158/0008-5472.CAN-05-2801)

[87] Jo M, Kim TH, Seol DW, Esplen JE, Dorko K, Billiar TR & Strom SC 2000 Apoptosis induced in normal human hepatocytes by tumor necrosis

factor-related apoptosis-inducing ligand. Nature Medicine **6** 564-567. (doi:10.1038/75045)

[88] Leverkus M, Neumann M, Mengling T, Rauch CT, Bröcker E-B, Krammer PH & Walczak H 2000 Regulation of Tumor Necrosis Factor-related Apoptosis-inducing Ligand Sensitivity in Primary and Transformed Human Keratinocytes. Cancer Research **60** 553-559.

[89] von Karstedt S, Conti A, Nobis M, Montinaro A, Hartwig T, Lemke J, Legler K, Annewanter F, Campbell AD, Taraborrelli L *et al.* 2015 Cancer cell-autonomous TRAIL-R signaling promotes KRAS-driven cancer progression, invasion, and metastasis. *Cancer Cell* **27** 561-573. (doi:10.1016/j.ccell.2015.02.014)

[90] Volkmann X, Fischer U, Bahr MJ, Ott M, Lehner F, Macfarlane M, Cohen GM, Manns MP, Schulze-Osthoff K & Bantel H 2007 Increased hepatotoxicity of tumor necrosis factor-related apoptosis-inducing ligand in diseased human liver. *Hepatology (Baltimore, Md.)* **46** 1498-1508. (doi:10.1002/hep.21846)

[91] Haselmann V, Kurz A, Bertsch U, Hübner S, Olempska–Müller M, Fritsch J, Häsler R, Pickl A, Fritsche H, Annewanter F *et al.* 2014 Nuclear Death Receptor TRAIL-R2 Inhibits Maturation of Let-7 and Promotes Proliferation of Pancreatic and Other Tumor Cells. *Gastroenterology* **146** 278-290. (doi:10.1053/j.gastro.2013.10.009)

[92] Lovric MM & Hawkins CJ 2010 TRAIL treatment provokes mutations in surviving cells. Oncogene **29** 5048-5060. (doi:10.1038/onc.2010.242)

[93] Infante JR, Dees EC, Olszanski AJ, Dhuria SV, Sen S, Cameron S & Cohen RB 2014 Phase I Dose-Escalation Study of LCL161, an Oral Inhibitor of Apoptosis Proteins Inhibitor, in Patients With Advanced Solid Tumors. Journal of Clinical Oncology **32** 3103-3110. (doi:10.1200/JCO.2013.52.3993)

[94] Bedoui S, Herold MJ & Strasser A 2020 Emerging connectivity of programmed cell death pathways and its physiological implications. *Nature Reviews.* Molecular Cell Biology **21** 678-695. (doi:10.1038/s41580-020-0270-8)

[95] Liebl MC & Hofmann TG 2019 Cell Fate Regulation upon DNA Damage: p53 Serine 46 Kinases Pave the Cell Death Road. BioEssays: News and Reviews in Molecular, Cellular and Developmental Biology **41** e1900127. (doi:10.1002/bies.201900127)

[96] Liu J-R, Wu Y-M, Xu G-M, Gao L-G, Ma Y, Shi X-M & Zhang G-J 2019 Low-temperature plasma induced melanoma apoptosis by triggering a p53/PIGs/caspase-dependent pathwayin vivoandin vitro. Journal of Physics D: Applied Physics **52** 315204. (doi: 10.1088/1361-6463/ab1ca9)

[97] Hafner A, Bulyk ML, Jambhekar A & Lahav G 2019 The multiple mechanisms that regulate p53 activity and cell fate. *Nature Reviews.* Molecular Cell Biology **20** 199-210. (doi:10.1038/s41580-019-0110-x)

[98] Kruse J-P & Gu W 2009 MSL2 promotes Mdm2-independent cytoplasmic localization of p53. *The Journal of Biological Chemistry* **284** 3250-3263. (doi:10.1074/jbc.M805658200)

[99] Di Stefano V, Rinaldo C, Sacchi A, Soddu S & D'Orazi G 2004 Homeodomain-interacting protein kinase-2 activity and p53 phosphorylation are critical events for cisplatin-mediated apoptosis. *Experimental Cell Research* **293** 311-320. (doi:10.1016/j.yexcr.2003.09.032)

[100] Ichwan SJA, Yamada S, Sumrejkanchanakij P,

Ibrahim-Auerkari E, Eto K & Ikeda M-A 2006 Defect in serine 46 phosphorylation of p53 contributes to acquisition of p53 resistance in oral squamous cell carcinoma cells. Oncogene **25** 1216-1224. (doi:10.1038/sj.onc.1209158)

[101] Okamura S, Arakawa H, Tanaka T, Nakanishi H, Ng CC, Taya Y, Monden M & Nakamura Y 2001 p53DINP1, a p53-inducible gene, regulates p53-dependent apoptosis. Molecular Cell **8** 85-94. (doi:10.1016/s1097-2765(01)00284-2)

[102] D'Orazi G, Cecchinelli B, Bruno T, Manni I, Higashimoto Y, Saito S, Gostissa M, Coen S, Marchetti A, Del Sal G *et al.* 2002 Homeodomain-interacting protein kinase-2 phosphorylates p53 at Ser 46 and mediates apoptosis. *Nature Cell Biology* **4** 11-19. (doi:10.1038/ncb714)

[103] Hofmann TG, Möller A, Sirma H, Zentgraf H, Taya Y, Dröge W, Will H & Schmitz ML 2002 Regulation of p53 activity by its interaction with homeodomain-interacting protein kinase-2. Nature Cell Biology **4** 1-10. (doi:10.1038/ncb715)

[104] Taira N, Nihira K, Yamaguchi T, Miki Y & Yoshida K 2007 DYRK2 is targeted to the nucleus and controls p53 via Ser46 phosphorylation in the apoptotic response to DNA damage. Molecular Cell **25** 725-738. (doi:10.1016/j.molcel.2007.02.007)

[105] Pfeffer CM & Singh ATK 2018 Apoptosis: A Target for Anticancer Therapy. International Journal of Molecular Sciences **19**. (doi:10.3390/ijms19020448)

[106] La Porta CAM & Zapperi S 2017 Complexity in cancer stem cells and tumor evolution: Toward precision medicine. *Seminars in Cancer Biology* **44** 3-9. (doi:10.1016/j.semcancer.2017.02.007)

[107] Thoma C 2017 Bladder cancer: The promise of liquid biopsy ctDNA analysis. *Nature Reviews.* Urology **14** 580-581. (doi:10.1038/nrurol.2017.138)

[108] Labi V & Erlacher M 2015 How cell death shapes cancer. Cell Death & Disease **6** e1675–e1675. (doi:10.1038/cddis.2015.20)

[109] Fogarty CE & Bergmann A 2017 Killers creating new life: caspases drive apoptosis-induced proliferation in tissue repair and disease. Cell Death and Differentiation **24** 1390-1400. (doi:10.1038/cdd.2017.47)

[110] Donato AL, Huang Q, Liu X, Li F, Zimmerman MA & Li C-Y 2014 Caspase 3 promotes surviving melanoma tumor cell growth after cytotoxic therapy. The Journal of Investigative Dermatology **134** 1686-1692. (doi:10.1038/jid.2014.18)

[111] Huang Q, Li F, Liu X, Li W, Shi W, Liu F-F, O'Sullivan B, He Z, Peng Y, Tan A-C *et al.* 2011 Caspase 3–mediated stimulation of tumor cell repopulation during cancer radiotherapy. Nature Medicine **17** 860-866. (doi:10.1038/nm.2385)

[112] Huang K, O'Neill KL, Li J, Zhou W, Han N, Pang X, Wu W, Struble L, Borgstahl G, Liu Z *et al.* 2019b BH3-only proteins target BCL-xL/MCL-1, not BAX/BAK, to initiate apoptosis. Cell Research **29** 942-952. (doi:10.1038/s41422-019-0231-y)

[113] Segawa K & Nagata S 2015 An Apoptotic 'Eat Me' Signal: Phosphatidylserine Exposure. Trends in Cell Biology **25** 639-650. (doi:10.1016/j.tcb.2015.08.003)

[114] Wang H, Cai J, Du S, Guo Z, Xin B, Wang J, Wei W & Shen X 2017 Fractalkine/CX3CR1 induces apoptosis resistance and proliferation through the activation of the AKT/NF-κB cascade in pancreatic cancer cells. Cell Biochemistry and Function **35** 315-326. (doi:10.1002/cbf.3278)

[115] Werfel TA & Cook RS 2018 Efferocytosis in the tumor micro-environment. Seminars in Immuno-pathology **40** 545-554. (doi:10.1007/s00281-018-0698-5)

[116] Ford CA, Petrova S, Pound JD, Voss JJLP, Melville L, Paterson M, Farnworth SL, Gallimore AM, Cuff S, Wheadon H *et al.* 2015 Oncogenic properties of apoptotic tumor cells in aggressive B cell lymphoma. *Current Biology: CB* **25** 577-588. (doi:10.1016/j.cub.2014.12.059)

[117] Fulda S 2009b Tumor resistance to apoptosis. International Journal of Cancer **124** 511-515. (doi:https://doi.org/10.1002/ijc.24064)

[118] Plati J, Bucur O & Khosravi-Far R 2008 Dysregulation of apoptotic signaling in cancer: Molecular mechanisms and therapeutic opportunities. Journal of Cellular Biochemistry **104** 1124-1149. (doi:https://doi.org/10.1002/jcb.21707)

[119] Gacche RN & Assaraf YG 2018 Redundant angiogenic signaling and tumor drug resistance. Drug Resistance Updates **36** 47-76. (doi:10.1016/j.drup.2018.01.002)

[120] Gottesman MM, Fojo T & Bates SE 2002 Multidrug resistance in cancer: role of ATP-dependent transporters. Nature Reviews. Cancer **2** 48-58. (doi:10.1038/nrc706)10.1038/nrc706)

[121] Leonetti A, Sharma S, Minari R, Perego P, Giovannetti E & Tiseo M 2019 Resistance mechanisms to osimertinib in EGFR-mutated non-small cell lung cancer. British Journal of Cancer **121** 725-737. (doi:10.1038/s41416-019-0573-8)

[122] Robey RW, Pluchino KM, Hall MD, Fojo AT, Bates SE & Gottesman MM 2018 Revisiting the role of ABC transporters in multidrug-resistant cancer. Nature Reviews Cancer **18** 452-464. (doi:10.1038/s41568-018-0005-8)

[123] Assaraf YG, Brozovic A, Gonçalves AC, Jurkovicova D, Linē A, Machuqueiro M, Saponara S, Sarmento-Ribeiro AB, Xavier CPR & Vasconcelos MH 2019 The multi-factorial nature of clinical multidrug resistance in cancer. Drug Resistance Updates: Reviews and Commentaries in Anti-microbial and Anticancer Chemotherapy **46** 100645. (doi:10.1016/j.drup.2019.100645)

[124] Goler-Baron V & Assaraf YG 2011 Structure and function of ABCG2-rich extracellular vesicles mediating multidrug resistance. PloS One **6** e16007. (doi:10.1371/journal.pone.0016007)

[125] Mansoori B, Mohammadi A, Davudian S, Shirjang S & Baradaran B 2017 The Different Mechanisms of Cancer Drug Resistance: A Brief Review. *Advanced Pharmaceutical Bulletin*. (doi:10.15171/apb.2017.041)

[126] Sherlach KS & Roepe PD 2014 Drug resistance associated membrane proteins. Frontiers in Physiology **5** 108. (doi:10.3389/fphys.2014.00108)

[127] Adams JM & Cory S 1998 The Bcl-2 protein family: arbiters of cell survival. *Science (New York, N.Y.)* **281** 1322-1326. (doi:10.1126/science.281.5381.1322)

[128] Kale J, Osterlund EJ & Andrews DW 2018 BCL-2 family proteins: changing partners in the dance towards death. Cell Death & Differentiation **25** 65-80. (doi:10.1038/cdd.2017.186)

[129] Lomonosova E & Chinnadurai G 2008 BH3-only proteins in apoptosis and beyond: an overview. Oncogene **27 Suppl 1** S2-19. (doi:10.1038/onc.2009.39)

[130] Delbridge ARD, Pang SHM, Vandenberg CJ, Grabow S, Aubrey BJ, Tai L, Herold MJ & Strasser A 2016 RAG-induced DNA lesions activate

proapoptotic BIM to suppress lymphomagenesis in p53-deficient mice. The Journal of Experimental Medicine **213** 2039-2048. (doi:10.1084/jem.20150477)

[131] Glab JA, Doerflinger M & Puthalakath H 2017 BH3-only proteins: the thorny end of the ER stress response. Cell Death & Disease **8** e2889. (doi:10.1038/cddis.2017.283)

[132] Huang DC & Strasser A 2000 BH3-Only proteins-essential initiators of apoptotic cell death. Cell **103** 839-842. (doi:10.1016/s0092-8674(00)00187-2)

[133] Suvarna V, Singh V & Murahari M 2019 Current overview on the clinical update of Bcl-2 anti-apoptotic inhibitors for cancer therapy. European Journal of Pharmacology **862** 172655. (doi:10.1016/j.ejphar.2019.172655)

[134] Maji S, Panda S, Samal SK, Shriwas O, Rath R, Pellecchia M, Emdad L, Das SK, Fisher PB & Dash R 2018 Chapter Three - Bcl-2 Antiapoptotic Family Proteins and Chemoresistance in Cancer. In *Advances in Cancer Research*, pp 37-75. Eds KD Tew & PB Fisher. Academic Press. (doi:10.1016/bs.acr.2017.11.001)

[135] Quinn BA, Dash R, Azab B, Sarkar S, Das SK, Kumar S, Oyesanya RA, Dasgupta S, Dent P, Grant S *et al.* 2011 Targeting Mcl-1 for the therapy of cancer. *Expert Opinion on Investigational Drugs* **20** 1397-1411. (doi:10.1517/13543784.2011.609167)

[136] Krajewska M, Moss SF, Krajewski S, Song K, Holt PR & Reed JC 1996 Elevated expression of Bcl-X and reduced Bak in primary colorectal adenocarcinomas. Cancer Research **56** 2422-2427.

[137] Minn AJ, Rudin CM, Boise LH & Thompson CB 1995 Expression of bcl-xL can confer a multidrug resistance phenotype. Blood **86** 1903-1910.

[138] Williams J, Lucas PC, Griffith KA, Choi M, Fogoros S, Hu YY & Liu JR 2005 Expression of Bcl-xL in ovarian carcinoma is associated with chemoresistance and recurrent disease. Gynecologic Oncology **96** 287-295. (doi:10.1016/j.ygyno.2004.10.026)

[139] Yang D, Chen M-B, Wang L-Q, Yang L, Liu C-Y & Lu P-H 2013 Bcl-2 expression predicts sensitivity to chemotherapy in breast cancer: a systematic review and meta-analysis. Journal of Experimental & Clinical Cancer Research **32** 105. (doi:10.1186/1756-9966-32-105)

[140] Pepper C, Hoy T & Bentley DP 1997 Bcl-2/Bax ratios in chronic lymphocytic leukaemia and their correlation with in vitro apoptosis and clinical resistance. British Journal of Cancer **76** 935-938. (doi:10.1038/bjc.1997.487)

[141] Czabotar PE, Lessene G, Strasser A & Adams JM 2014 Control of apoptosis by the BCL-2 protein family: implications for physiology and therapy. Nature Reviews. Molecular Cell Biology **15** 49-63. (doi:10.1038/nrm3722)

[142] Jeng L-B, Kumar Velmurugan B, Chen M-C, Hsu H-H, Ho T-J, Day C-H, Lin Y-M, Padma VV, Tu C-C & Huang C-Y 2018 Fisetin mediated apoptotic cell death in parental and Oxaliplatin/irinotecan resistant colorectal cancer cells in vitro and in vivo. Journal of Cellular Physiology **233** 7134-7142. (doi:10.1002/jcp.26532)

[143] Chen W, Li N, Chen T, Han Y, Li C, Wang Y, He W, Zhang L, Wan T & Cao X 2005 The lysosome-associated apoptosis-inducing protein containing the pleckstrin homology (PH) and FYVE domains (LAPF), representative of a novel family of PH and FYVE domain-containing proteins, induces caspase-independent apoptosis via the lysosomal-mitochondrial pathway. The Journal of Biological Chemistry **280** 40985-40995. (doi:10.1074/jbc.M502190200)

[144] Roberts AW, Davids MS, Pagel JM, Kahl BS, Puvvada SD, Gerecitano JF, Kipps TJ, Anderson MA, Brown JR, Gressick L *et al.* 2016 Targeting BCL2 with Venetoclax in Relapsed Chronic Lymphocytic Leukemia. *New England Journal of Medicine* **374** 311-322. (doi:10.1056/NEJMoa1513257)

[145] Wilson WH, O'Connor OA, Czuczman MS, LaCasce AS, Gerecitano JF, Leonard JP, Tulpule A, Dunleavy K, Xiong H, Chiu Y-L *et al.* 2010 Navitoclax, a targeted high-affinity inhibitor of BCL-2, in lymphoid malignancies: a phase 1 dose-escalation study of safety, pharmacokinetics, pharmacodynamics, and antitumour activity. The Lancet. Oncology **11** 1149-1159. (doi:10.1016/S1470-2045(10)70261-8)

[146] Mason KD, Vandenberg CJ, Scott CL, Wei AH, Cory S, Huang DCS & Roberts AW 2008 In vivo efficacy of the Bcl-2 antagonist ABT-737 against aggressive Myc-driven lymphomas. *Proceedings of the National Academy of Sciences of the United States of America* **105** 17961-17966. (doi:10.1073/pnas.0809957105)

[147] Sasi BK, Martines C, Xerxa E, Porro F, Kalkan H, Fazio R, Turkalj S, Bojnik E, Pyrzynska B, Stachura J *et al.* 2019 Inhibition of SYK or BTK augments venetoclax sensitivity in SHP1-negative/BCL-2-positive diffuse large B-cell lymphoma. Leukemia **33** 2416-2428. (doi:10.1038/s41375-019-0442-8)

[148] Han B, Park D, Li R, Xie M, Owonikoko TK, Zhang G, Sica GL, Ding C, Zhou J, Magis AT *et al.* 2015 Small-Molecule Bcl2 BH4 Antagonist for Lung Cancer Therapy. Cancer Cell **27** 852-863. (doi:10.1016/j.ccell.2015.04.010)

[149] Dexheimer TS, Sun D & Hurley LH 2006 Deconvoluting the structural and drug-recognition complexity of the G-quadruplex-forming region upstream of the bcl-2 P1 promoter. Journal of the American Chemical Society **128** 5404-5415. (doi:10.1021/ja0563861)

[150] Sun H, Xiang J, Shi Y, Yang Q, Guan A, Li Q, Yu L, Shang Q, Zhang H, Tang Y *et al.* 2014 A newly identified G-quadruplex as a potential target regulating Bcl-2 expression. *Biochimica et Biophysica Acta (BBA) - General Subjects* **1840** 3052-3057. (doi:10.1016/j.bbagen.2014.07.014)

[151] Follis AV, Chipuk JE, Fisher JC, Yun M-K, Grace CR, Nourse A, Baran K, Ou L, Min L, White SW *et al.* 2013 PUMA binding induces partial unfolding within BCL-xL to disrupt p53 binding and promote apoptosis. *Nature Chemical Biology* **9** 163-168. (doi:10.1038/nchembio.1166)

[152] Hao Q, Chen J, Liao J, Huang Y, Larisch S, Zeng SX, Lu H & Zhou X 2020 p53 induces ARTS to promote mitochondrial apoptosis. BioRxiv 2020.05.14.096982. (doi:10.1101/2020.05.14.096982)

[153] Adams JM & Cory S 2018 The BCL-2 arbiters of apoptosis and their growing role as cancer targets. Cell Death and Differentiation **25** 27-36. (doi:10.1038/cdd.2017.161)

[154] Müschen M & Beckmann MW 2000 CD95 ligand expression as a criterion of malignant transformation in breast cancer. The Journal of Pathology **191** 468-470. (doi:10.1002/1096-9896(200008)191:4<468::AID-PATH647>3.0.CO;2-J)

[155] Tourneur L, Buzyn A & Chiocchia G 2005 FADD adaptor in cancer. *Medical Immunology (London, England)* **4** 1. (doi:10.1186/1476-9433-4-1)

[156] Müschen M, Rajewsky K, Krönke M & Küppers R 2002 The origin of CD95-gene mutations in B-cell lymphoma. Trends in Immunology **23**

75-80. (doi:10.1016/s1471-4906(01)02115-9)

[157] Tourneur L, Delluc S, Lévy V, Valensi F, Radford-Weiss I, Legrand O, Vargaftig J, Boix C, Macintyre EA, Varet B *et al.* 2004 Absence or low expression of fas-associated protein with death domain in acute myeloid leukemia cells predicts resistance to chemotherapy and poor outcome. Cancer Research **64** 8101-8108. (doi:10.1158/0008-5472.CAN-04-2361)

[158] Shivapurkar N, Toyooka S, Eby MT, Huang CX, Sathyanarayana UG, Cunningham HT, Reddy JL, Brambilla E, Takahashi T, Minna JD *et al.* 2002 Differential inactivation of caspase-8 in lung cancers. *Cancer Biology and Therapy* **1** 65-69. (doi:10.4161/cbt.1.1.43)

[159] Teitz T, Wei T, Valentine MB, Vanin EF, Grenet J, Valentine VA, Behm FG, Look AT, Lahti JM & Kidd VJ 2000 Caspase 8 is deleted or silenced preferentially in childhood neuroblastomas with amplification of MYCN. Nature Medicine **6** 529-535. (doi:10.1038/75007)

[160] Zhang L, Zhu H, Teraishi F, Davis JJ, Guo W, Fan Z & Fang B 2005 Accelerated Degradation of Caspase-8 Protein Correlates with TRAIL Resistance in a DLD1 Human Colon Cancer Cell Line. Neoplasia 7 594-602. (doi:10.1593/neo.04688)

[161] Almasan A & Ashkenazi A 2003 Apo2L/TRAIL: apoptosis signaling, biology, and potential for cancer therapy. Cytokine & Growth Factor Reviews **14** 337-348. (doi:10.1016/S1359-6101(03)00029-7)

[162] Amarante-Mendes GP & Griffith TS 2015 Therapeutic applications of TRAIL receptor agonists in cancer and beyond. Pharmacology & Therapeutics **155** 117-131. (doi:10.1016/j.pharmthera.2015.09.001)

[163] Bucur O, Gaidos G, Yatawara A, Pennarun B, Rupasinghe C, Roux J, Andrei S, Guo B, Panaitiu A, Pellegrini M *et al.* 2015 A novel caspase 8 selective small molecule potentiates TRAIL-induced cell death. Scientific Reports **5** 9893. (doi:10.1038/srep09893)

[164] Ashkenazi A, Pai RC, Fong S, Leung S, Lawrence DA, Marsters SA, Blackie C, Chang L, McMurtrey AE, Hebert A *et al.* 1999 Safety and antitumor activity of recombinant soluble Apo2 ligand. *The Journal of Clinical Investigation* **104** 155-162. (doi:10.1172/JCI6926)

[165] LeBlanc HN & Ashkenazi A 2003 Apo2L/TRAIL and its death and decoy receptors. Cell Death and Differentiation **10** 66-75. (doi:10.1038/sj.cdd.4401187)

[166] Sasaki H, Sheng Y, Kotsuji F & Tsang BK 2000 Down-Regulation of X-linked Inhibitor of Apoptosis Protein Induces Apoptosis in Chemoresistant Human Ovarian Cancer Cells. Cancer Research **60** 5659-5666.

[167] Holcik M, Yeh C, Korneluk RG & Chow T 2000 Translational upregulation of X-linked inhibitor of apoptosis (XIAP) increases resistance to radiation induced cell death. Oncogene **19** 4174-4177. (doi:10.1038/sj.onc.1203765)

[168] Amantana A, London CA, Iversen PL & Devi GR 2004 X-linked inhibitor of apoptosis protein inhibition induces apoptosis and enhances chemotherapy sensitivity in human prostate cancer cells. Molecular Cancer Therapeutics **3** 699-707.

[169] Fulda S 2015 Promises and Challenges of Smac Mimetics as Cancer Therapeutics. Clinical Cancer Research: An Official Journal of the American Association for Cancer Research **21** 5030-5036. (doi:10.1158/1078-0432.CCR-15-0365)

Chapter 3

Recent Advancements in Apoptosis-Based Therapeutic Approaches for Cancer Targeting

Mehmet Evren Okur, Panoraia I. Siafaka, Merve Tutar and Yusuf Tutar

Abstract

Apoptosis, known as programmed cell death, has been considered a potent target for the pharmacy industry. The scientific community has actively participated to research which evaluate active molecules for possible inhibition or induction of apoptosis. Nanocarriers especially for cancer targeting are widely found through literature; they mainly based on inorganic, lipid or polymer nanoparticles which incorporate anticancer drugs. Another important and innovative category of anticancer agents is that of microRNAs. In this chapter, a discussion about the most recent applications of apoptosis-based agents mainly focusing on cancer target is done.

Keywords: apoptotic agents, anticancer agents, microRNAs

1. Introduction

In the last decades, a huge advancement in the development of novel drug targets and drug delivery systems for inhibiting or inducing apoptosis has been done. Apoptosis, the programmed cell death plays a major role in cellular homeostasis, normal development, and clearance of cells. Non-programmed cell death can take place by various external factors, such as infection, toxins, and physical injury [1]. The dysregulation of apoptosis has been related with the pathogenesis of numerous diseases such as degenerative, autoimmune, and cardiovascular as well as have been associated with tumorigenesis. It has been reported that apoptosis is reduced on pathological disorders such as malignant neoplasm and autoimmune diseases while it is raised in inflammatory and neurodegenerative disorders as well as ischemic diseases, *i.e.* myocardial infarction, liver ischemia, stroke [2]. Apoptosis can induce cancer formation while cancer cells may elude apoptosis via the downregulation or blockage of apoptosis signaling pathways [3]. Similarly, cell death modalities (apoptosis, necrosis, and autophagy) have been linked with cardiovascular, autoimmune, and neurodegenerative diseases. In example, it has been reported that the apoptosis is impaired by various factors (*i.e.* caspases, amyloid β, tumor necrosis factor-α, amyloid precursor protein intracellular C-terminal domain, etc.) leading to Alzheimer's disease and other neurodegenerative diseases (such as Huntington's disease) [4–6]. Moreover, cardiovascular disorders such as myocardial infarction, diabetic cardiomyopathy, ischemic cardiomyocyte, and congestive heart failure have been associated with cell death modes in cardiac myocytes [7].

Identification of key players in cellular apoptosis regulation as B-cell lymphoma 2 (BCL-2) proteins, caspases, etc. has proofed that targeting apoptosis can lead to outstanding management of various diseases, especially cancer [8, 9].

2. Nanocarriers inducing apoptosis

In general, pharmacological approaches related to apoptosis can be categorized as inhibiting and inducing apoptosis molecules. Inhibitors of apoptosis (IAPs) are proteins important for maintaining apoptosis; IAPs have been identified in tumors and thus therapies targeting them have raised the research interest. It is well known that cancer cells are more resistant to apoptotic cell death, and high dosages of drugs are needed to eliminate them [10]. Consequently, the usage of apoptotic pathways via IAP antagonism can act as a promising alternative therapeutic choice for cancer management, limiting the apoptotic effect on cancerous cells [11]. In case of cardiovascular disorders, inhibiting myocardial apoptosis has been recognized as a therapeutic target. In the past, caspase inhibitors [12] and control of apoptosis of cardiovascular fibroblasts/vascular smooth muscle cells by p65 nuclear factors NF-κB and B-cell lymphoma-extra-large (Bcl-xL) antisense oligonucleotides or p53 overexpression [13] have been reported as potent drug targets. A recent study [14] revealed that the upregulation of miR-29b-3p (miRNAs), can protect cardiomyocytes against hypoxia-induced injury through downregulation of TNF receptor-associated factor 5 which can be an important therapeutic alternative for acute myocardial infarction. BH3 interacting-domain death agonist [BID (belongs to BCL-2 family)] inhibitors have been recognized as key factors in the apoptotic pathway mediating cytochrome C and Smac/DIABLO from mitochondria, resulting in caspase activation and cell apoptosis. The inhibition of BID by pharmacological agents can offer a promising therapeutic choice for diseases implicated by pathological cell death from BID involvement [2]. Becattini *et al.* developed various 4-phenylsulfanyl-phenylamine derivatives that are capable of binding on the surface of BID resulting to inhibition of tBid-induced SMAC release, caspase-3 activation, and cell death [15]. Tumor necrosis factor-like weak inducer of apoptosis (TWEAK) inhibitors have been also reported to induce apoptosis; tumor necrosis factor (TNF) has been linked with cerebral ischemia, atherosclerosis, rheumatoid arthritis, etc. The TWEAK-Fibroblast growth factor-inducible 14 (FN-14) has been identified as potent drug target for the aforementioned diseases [16]. Cytochrome C inhibitors, as Minocycline and methazolamide have shown inhibition of apoptosis. Minocycline revealed inhibition of apoptosis via attenuation of TNF-alpha expression following iNOS/NO induction by lipopolysaccharide in neuron/glia co-cultures [17]. Tian et al. showed that minocycline can inhibit sevoflurane-induced apoptosis [18]. Methazolamide, a carbonic anhydrase has shown prevention of the amyloid β-mediated onset of apoptosis in the mouse brain [19]. Therefore, various apoptotic inhibitors have been developed; however, a great effort is much needed.

A special reference on the drug carriers involved as chemotherapeutic systems inducing apoptosis of cancerous cells is followed due to its importance. Cancer is the second leading cause of death globally [20] and despite the great effort being done by researchers worldwide, its management remains inefficient. Chemotherapy, the most conventional treatment option can lead to serious adverse effects since it can induce the cell death of healthy tissues as well as the cancerous ones [21]. In most cases, the current therapeutic systems involve the use of nanoparticles based on polymeric and inorganic carriers and a combination of them. Nanotechnology-based carriers are of great interest due to their limited size, improved penetration, and functionalization potential resulting to targeting efficiency [22, 23]. The therapeutic outcome of cancer management strategies mainly depends to the ability of the molecule to induce apoptosis via targeting the overexpressed anti-apoptotic proteins or stimulation of the expression of pro-apoptotic molecules. Nonetheless, cancer cells seem to resist the chemotherapy and apoptosis leading to increased survival rate and the possibility of metastasis. Thus, the combination of drugs or novel strategies involving innovative drug carriers can potentially overcome the chemoresistance [24, 25]. Solid lipid nanoparticles belong to lipid-based carriers and have been designed as potent drug delivery systems for various diseases. As an example, letrozole, a known cytotoxic agent for hormone-dependent breast cancer management was impregnated to folic acid-modified solid lipid nanoparticles in order to induce apoptotic cell death. The modification with folic acid can lead to enhanced targeting efficiency since breast cancer cells overexpressed folate receptors. Inducing of apoptosis performed by employing caspase-3 activity and TUNEL assays. The incorporation of the drug into the folic acid decorated carriers led to *in vitro* cytotoxicity against MCF-7 cancer cells but they were not cytotoxic to MCF-10A normal cells revealing great biocompatibility. It was concluded that the mechanism of cell death was apoptosis based [26]. Similarly, solid lipid nanoparticles were modified with chitosan-coated-trans-resveratrol and ferulic acid and further decorated with folic acid. The developed nano-formulations demonstrated great stability and improved cytotoxicity in the colon cancer cells which led to apoptotic cell death. Thus, the solid lipid nanoparticles can be used for anticancer therapy [27]. Folate modified hydroxyapatite nanorods were used as a matrix for doxorubicin, an anticancer agent, loading. The results showed cytotoxicity against MCF-7 cells while western blot assays revealed that the developed nanocarriers can improve the expression of Bax (a pro-apoptotic protein) and decrease the expression of Bcl-2. Finally, the nanorods improved mitochondrial cytochrome C leakage and activate an apoptotic cell death [28]. In similar skeptic, folic acid conjugated chitosan nanoparticles were able to incorporate the cytotoxic agent, ursolic acid. The targeting affinity improved the local concentration of the drug in cancerous cells MCF-7. According to the studies, the nanoparticles enter the lysosome, released from it while it was localized into mitochondria but not nuclei. Their prolonged retention in mitochondria led to the irreversible apoptosis in cancer cells owing to the overproduction of ROS and the destruction of the mitochondrial membrane. According to the mouse xenograft model, the nanoparticles can hamper breast cancer revealing promising characteristics and potent clinical efficacy [29]. Nano-formulations comprised of amine-functionalized and conjugated with folic acid mesoporous silica nanoparticles were loaded with curcumin, quercetin, and colchicines, known as anticancer prodrugs. The folate decorated nanoparticles incorporating curcumin revealed greater cellular uptake, prolonged intracellular release, and cytotoxicity. It was also reported that the apoptotic cell death was induced through specific signaling molecular pathways (caspase-3, H_2O_2, c-MET, and MCL-1), providing great [30]. In the last years the combination of inorganic nanoparticles, such as mesoporous silica nanoparticles and polymeric materials has been winning the race for effective cancer management. In example, mesoporous silica nanoparticles loaded with topotecan and externally modified with poly (acrylic acid) co-synthesized with chitosan, were investigated for their efficacy against triple-negative breast cancer (MDA-MB-231) and multi-drug resistant MCF-7 cells. The external layer is conjugated also with quercetin as a second drug. Moreover, arginine-glycine-aspartic acid peptide was conjugated on the nanoparticles inducing their uptake from the cancer cells via integrin receptor-mediated endocytosis. It was concluded that the system promoted molecular activation and cell death [31]. Folate modified liposomes loaded with bleomycin were prepared via film hydration and studied for their anticancer activity and apoptosis induction. According to the MTT assay, the nano-liposomal formulations showed greater efficacy in human cervix carcinoma HeLa, and human breast carcinoma MCF-7 cells. Due to the modification of formulations with folic acid, an improved uptake by HeLa cells was confirmed. Additionally, the folate nano-liposomes with bleomycin effectively promoted apoptotic cell death as well as a cell-cycle arrest in HeLa cells especially at the G2/M phase [32]. Another system composed of modified chitosan nanoparticles and methotrexate was investigated as a potent anticancer system. Two molecules, L-cysteine and folic acid were conjugated to chitosan; the decorated nanoparticles when studied in a reducing environment similar to tumor cells, released the drug as desired. Moreover, the nanoparticles induced anticancer activity on HeLa cells in a dose and time-dependent manner while they demonstrated selective cellular uptake [33]. Another research involved the incorporation of doxorubicin into folic acid-modified lactoglobulin nanoparticles and studied for their anticancer potential against MCF-7 and MDA-MB-231, BC and triple-negative BC cells. It was revealed an important inhibition of cell proliferation and promotion of apoptosis [34].

DR5 which belongs to the TNFR family has been proposed as a potential target for cancer. An interesting study investigated the application of poly(ethylene glycol) decorated poly(lactic-co-glycolic acid) nanoparticles as potent anticancer carriers. The nanoparticles were further conjugated with DR5-specific antibody conatumumab and impregnated with camptothecin. The stealth nanocarriers promoted pro-apoptotic effects of the platform *in vivo* using HCT116 adenocarcinoma xenografts [35]. Another research evaluated the development of nanoparticles comprised from copolymers between poly(ε-caprolactone)-PCL and poly(ethylene glycol)-PEG as well as PEG, PCL, and poly(lactic acid)-PLA as potent carriers of auraptene. The triblock PCL-PEG-PCL and pentablock PLA-PCL-PEG-PCL-PLA copolymers were formulated on nanoparticles and examined for their characteristics. According to the results, the PLA-PCL-PEG-PCL-PLA nanoparticles showed enhanced cellular uptake as well as cytotoxicity. In further, the nano-formulations incorporating auraptene promoted the apoptotic cell death on HT-29 colon cancer cells. The real-time PCR revealed as apoptosis marker the Bax /Bcl2 expression ratio which was increased in the case of pentablock nanoparticles [36]. Nanocarriers based on mPEGylated Dendron conjugated with glycylphenylalanylleucylglycine tetra-peptide spacer and doxorubicin were studied for their efficacy against multidrug resistance of cancer chemotherapy. The *in vitro* studies revealed that the nanoparticles can accumulate in the nuclei of MCF-7/ADR cells and they are potentially cytotoxic leading to apoptosis. Moreover, the nanoparticles showed enhanced therapeutic efficiency against multidrug resistance xenograft tumors and thus they can be applied as potent anticancer carriers that can initiate the lysosomal apoptosis pathway [37]. Nano-formulations based on triphenylphosphine, Pluronic F127, and hyaluronic acid, formulated on nanomicelles able to incorporate paclitaxel, anticancer drug. The nanosystem showed inhibition of A549/ADR cells. Moreover, the nanomicelles entered acidic lysosomes through macropinocytosis, and accumulate in the mitochondria over a day, in A549/ADR cells. The nanomicelles induced the permeabilization of the mitochondrial outer membrane via hindering anti-apoptotic Bcl-2, resulting in the release of cytochrome C as well as caspase-3 and caspase-9 activation. Accordingly, when the nanoformulations studied in A549/ADR xenograft tumor model and a drug-resistant breast cancer mice model with lung metastasis, demonstrated promising cancer targeting and desirable anticancer efficiency [38]. A promising way to initiate cancer cell apoptosis is to targeted deliver cytochrome c, which can mediate apoptotic cell death if released from the mitochondria to the cytoplasm. Thus, an innovative nanosystem based on Cytochrome c was developed in order to promote the apoptotic death of cancer cells when is delivered. The nanocarriers of cytochrome c were modified with poly (lactic-co-glycolic) acid-SH via the

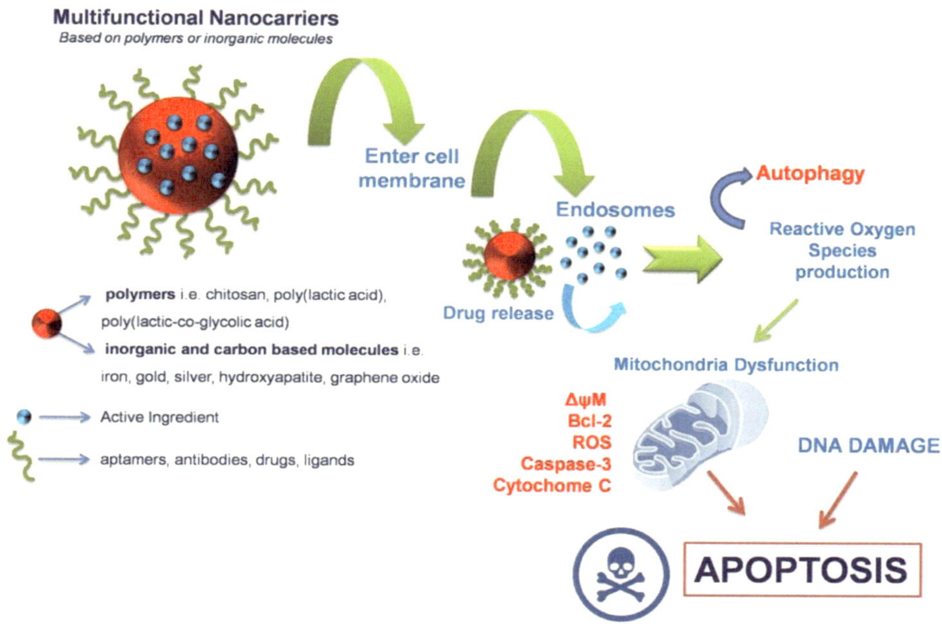

Figure 1.
A summary of multifunctional nanocarriers inducing apoptosis.

linker succinimidyl 3-(2-pyridyldithio) propionate so as to prevent the degradation possibility. The nanoformulations when incubated with HeLa cells exhibited cytotoxicity and the promotion of apoptosis in HeLa cells was also demonstrated [39]. Carbon dots have been widely studied as anticancer systems, *i.e.* carbon dots developed via ultrasonication of sucrose were modified with Gemcitabine and studied for their activity on MCF-7 and HeLa cell lines. The cytocompatible nano-formulations expressed desirable bioactivity and cytotoxicity against the cell lines. Moreover, the nano-formulations did not affect the healthy cells in such extent as they acted against cancerous cells. The most important outcome of the study is that the Gemcitabine conjugated carbon dots promoted early and late apoptotic cell deaths in the MCF-7 and HeLa cancer cell lines [40]. Graphene oxide modified with GE11 peptide was fabricated in order to efficiently targeted deliver oridonin on cancerous cells. The modified graphene oxide showed improved cellular uptake in KYSE-30 and EC109 esophageal cells compared to healthy cells. The oridonin loaded carriers were able to accumulate into lysosomes while they desirably hinder the viability of cancer cells. In addition, the developed system promoted apoptosis of the aforementioned cancerous cell lines [41]. **Figure 1** summarizes the possible mechanisms involving when nanocarriers are delivered to cells and induce apoptosis.

Several active compounds can be used for therapeutic purposes but miRNAs may also be employed for this purpose. The second part focuses on miRNAs as anticancer agents.

3. APOPTOTIC MicroRNAs (miRNA)

3.1 How miRNAs work to silence genes

To generate mature miRNAs, sequential steps are followed. First, RNA Polymerase II transcribes primary miRNAs (abbreviated as pri-miRNA). These molecules can be translated from both intergenic and intragenic regions [42]. Then in order to generate pre-miRNAs with hairpin precursors, pri-miRNAs are processed by Drosha, an RNase III enzyme and by Pasha, a dsRNA-binding protein [43]. After this step, pre-miRNAs get escorted out of the nucleus with the help of exportin 5 [44]. And the pre-miRNA gets further processed in cytoplasm by Dicer enzyme which is an RNase III endonuclease, this enzyme works to remove the hairpin loop and make a double stranded duplex miRNA [44]. Then this double stranded structure retains the active strand while passenger stand gets degraded. The active strand interacts with RNA-induced silencing complex (abbreviated as RISC) in order to function. RISC is a complex formed by multiple proteins with its key proteins being argonaute 2 (abbreviated as Ago2) and transactivation-responsive RNA-binding protein (abbreviated as TRBP), and it includes miRNA or siRNA in order to use them as a template [45]. With these RNA templates the complex is able to recognize their complementary mRNA. The characteristic features of the target gene for an effective binding can be; seed region, a target sequence that is conserved, 3′ untranslated region of miRNA available for binding but recent studies show the binding of a target may also happen in 5′ untranslated region, promoter regions or open reading frames [45]. The pairing of the miRNA template and its target mRNA differs between plant cells and animal cells. In the plant cells, the pairing is fully complementary between miRNA and mRNA. But in animal cells, this pairing is not fully complementary there are base mismatches even though this base pairing follows a pattern. But there is a small sequence with 2–8 nucleotides of length which is a perfect base pairing that is called seed region [46]. This region is a conserved heptametrical sequence that is always perfectly matched and it is mostly found towards the 5′ end of miRNA [47]. With the binding of mRNA and miRNA, downregulation is tried to be achieved and this can be achieved via enzymatic cleavage of the mRNA leading to its further destruction by the cell or blocking the translation by preventing ribosome subunit from binding to mRNA [48]. And the matching degree of target and miRNA plays a role in the decision of which mechanism will happen for downregulation to occur, if the target is fully complementary then cleavage of mRNA will happen but if it is not fully complementary stability alteration or repression of translation may occur [43].

3.2 miRNAs in apoptosis

miRNAs are known to have a regulatory effect on apoptosis via their regulation on both pro-apoptotic and anti-apoptotic genes. So, miRNAs can work to be both inhibitory and stimulatory depending on the miRNA and the cell context. Also, alteration of the expression of regulatory genes in the apoptotic process by miRNAs is not limited to one of the extrinsic and intrinsic pathways. And the effect of miRNAs can both be direct and indirect. For example, miR-21 is a miRNA that directly affects its target, inhibiting FasL in order to increase apoptosis but miR-130a is a miRNA that affects TRAIL resistance in order to effect other miRNAs that will eventually cause a change in apoptotic process [49].

For their indirect effects, miRNAs can be seen to function in both feedback and feedforward loops. Feedback loop effects can change depending on the cell context, miRNA and transcription factors as the regulators may have the same or opposite effects [50]. And, in feed-forward loops, transcription factors can be seen to regulate both the target gene and miRNA, which also regulates the transcription factor [50]. To regulate genes, miRNAs work together with transcription factors in a highly coordinated manner. Since they can show their effects on mRNAs after the transcription of said mRNA, they usually locate downstream to transcription factors [51].

In the intrinsic pathway, p53 and BCL-2 families play an important role. MiRNAs can alter their expression to regulate the intrinsic pathway. As miRNAs regulate the levels of p53, this tumor suppressor actually has an effect on the miRNAs as well by functioning to regulate miRNA expression and maturation [52]. For an example of p53 regulating miRNAs and how its mutation can cause a change, we can look at miR-16 and miR-143. Their processing is dependent on the interaction between p53 and Drosha complex so if there is a mutation in the DNA binding domain of p53, their processing cannot be achieved and cell proliferation will be suppressed [45]. Activation of p53 is found to be increasing the expression of 30 or more miRNAs including miRNAs like let7a, miR-34a and miR-15a/16 which are tumor suppressors [53]. BCL-2 is an anti-apoptotic protein that is generally overexpressed in tumors. Three pro-apoptotic miRNAs, miR-24, miR-195 and miR-365, work to down-regulate BCL2 expression via their binding to BCL-2 gene's 3′ untranslated region [53]. With this interaction, pro-apoptotic miRNAs lead to apoptosis. Extrinsic pathway is also regulated by miRNAs. Some miRNAs were found to regulate TRAIL-induced apoptosis directly and indirectly [43]. miR-221 and miR-222 can be an example of this regulation since they are found to have altering expressions between TRAIL resistant and sensitive cells, resistant cells being the ones with up-regulation of these miRNAs [43]. Another example can be miR-200c since it directly targets FAP-1, a phosphatase that works to inhibit apoptosis [43].

An example to miRNAs with effects not limited to one site is miR-21. We can observe its effects on both non-small cell lung carcinoma (NSCLC) and diffuse large B-cell lymphoma (DLBCL). In NSCL, miR-21 effects apoptosis via its inhibition on PI3K/Akt/NF-kB pathway and also it is found that miR-21 targets apoptosis-stimulating protein of p53 (ASPP2) which is a protein that functions in tumorigenesis [54]. And it was found that in early-stage samples of NSCL cells, miR-21 expression was increased when compared to the control [55]. The experiments revealed that in NSCL cells, miR-21 down-regulation led to the repression of EMT signaling pathway, cell migration and invasion, and miR-21 inhibition led to triggering of apoptosis [54]. Both *in vitro* and *in vivo*, miR-21 inhibited PI3K/Akt/NF-κB signaling pathway and promoted caspase-dependent pathway of apoptosis. MiR-21 also is known to have high expression levels in B-cell lymphoma. In DLBCL, its effect on apoptosis can be seen via regulation of phosphatase and tensin homolog (PTEN). The expression level of miR-21 in patient samples was found to be more than the healthy samples, and these levels were also negatively correlated with expression level of PTEN [14]. Other miRNAs that have an effect on PTEN are miR-130 family. This family of miRNAs which corresponds to miR-130b, miR-301a and miR-301b, are found to have high expression levels in bladder cancer samples compared to normal ones. Via their regulation upon PTEN they also regulate focal adhesion kinase (FAK) and Akt phosphorylation, and lead to cell migration and invasion increase in bladder cancer. Experiments showed that the inhibition of this family causes down-regulation of FAK and Akt phosphorylation and this effects cell migration and invasion negatively, so it can be said that they have an important role in the progression of bladder cancer [56].

As it can be seen from the examples, the effects of miRNAs are diversifiable depending on the gene they are affecting or their expression level. The alteration of their expression leads to interchangeable role of miRNAs as oncogenes or tumor suppressors. Generally, the miRNAs that are down-regulated in the cancer tissues are considered to be tumor suppressors, as pro-apoptotic miRNAs they work for apoptosis to happen. miR-7, miR335 and miR-608 are examples of this type of miRNAs since they target BCL-2 family, and miR-203 and miR-143 can be other examples as they target PKC family [57]. On the other hand, other miRNA examples can be seen as upregulated in the cancer tissues, as antiapoptotic miRNAs they induce apoptosis to allow uncontrollable proliferation. miR-197, miR-21 and miR-212 can be the examples of these kind of miRNAs [57].

Our current research based on developing target specific drug candidates over breast cancer cell lines. The array studies indicated that non-coding RNA hsa-miR-215 greatly enhance the inhibitor compound efficiency on MCF-7 and MDA-MB-231 breast cancer cell lines. The expression of hsa-miR-215 decreases in breast cancer cell lines compared to non-cancerous breast MCF10-A cell line. Over expressing this miRNA by transfecting into the cancerous cell lines drive the cells to apoptosis. Therefore, synergetic effect of the inhibitor compound along with hsa-miR-215 mimic augment anticancer treatment. A nanocarrier is being developed with hsa-miR-215 and inhibitor compound (patent pending). This formulation is designed for apoptosis-based therapeutic approach for breast cancer treatment. The utilization of carrier system along with miRNAs and inhibitor compounds introduced in this study for therapeutic purposes has the potential of clinical applications.

4. Future perspectives

Regulated cell death can be employed in cancer treatment since the apoptotic process can restrain survival of abnormal cells. miRNAs regulate distinct pathways and target the pathway's components. Cancer cells metabolic rate is higher and block apoptosis for survival strategy. Factors driving cancer cells to apoptosis can be used for treatment purposes. Recent advances in drug discovery and miRNA handling help these designs. Inhibitor compounds and miRNA-based therapeutics in oncology are promising and although face some challenges.

Acknowledgements

YT acknowledges project "Factors Providing Mitochondria and Cytosol Interaction in Apoptosis Regulation" (Grant # SBU 2017/041).

DOI: http://dx.doi.org/10.5772/intechopen.99202

Author details

Mehmet Evren Okur[1], Panoraia I. Siafaka[2,3], Merve Tutar[4] and Yusuf Tutar[5,6]*

1 Faculty of Pharmacy, Department of Pharmacology, University of Health Sciences, Istanbul, Turkey

2 Department of Chemistry, Aristotle University of Thessaloniki, Thessaloniki, Greece

3 KES College, Nicosia, Cyprus

4 Faculty of Science, Department of Molecular Biology and Genetics, Bilkent University, Ankara, Turkey

5 Hamidiye Faculty of Pharmacy, Division of Biochemistry, Department of Basic Pharmaceutical Sciences, University of Health Sciences, Istanbul, Turkey

6 Division of Molecular Oncology, Hamidiye Health Sciences Institutes, University of Health Sciences, Istanbul, Turkey

*Address all correspondence to: ytutar@outlook.com

References

[1] Yan G, Elbadawi M, Efferth T. Multiple cell death modalities and their key features (Review). World Acad Sci J 2020;2:39-48. doi:10.3892/wasj.2020.40

[2] Mohamed Naguib Abdelhafez E-S, Mohamed Naguib Abdelhafez Ali S, Ramadan Eisa Hassan M, Mohammed Abdel-Hakem A. Apoptotic Inhibitors as Therapeutic Targets for Cell Survival. Cytotox. - Defin. Identification, Cytotoxic Compd., IntechOpen; 2019. doi:10.5772/intechopen.85465

[3] Messmer MN, Snyder AG, Oberst A. Comparing the effects of different cell death programs in tumor progression and immunotherapy. Cell Death Differ 2019;26:115-129. doi:10.1038/s41418-018-0214-4

[4] Obulesu M, Lakshmi MJ. Apoptosis in Alzheimer's Disease: An Understanding of the Physiology, Pathology and Therapeutic Avenues. Neurochem Res 2014;39:2301-2312. doi:10.1007/s11064-014-1454-4

[5] Guo T, Zhang D, Zeng Y, Huang TY, Xu H, Zhao Y. Molecular and cellular mechanisms underlying the pathogenesis of Alzheimer's disease. Mol Neurodegener 2020;15:40. doi:10.1186/s13024-020-00391-7

[6] Bano D, Zanetti F, Mende Y, Nicotera P. Neurodegenerative processes in Huntington's disease. Cell Death Dis 2011;2:e228–e228. doi:10.1038/cddis.2011.112

[7] Lee Y, Gustafsson ÅB. Role of apoptosis in cardiovascular disease. Apoptosis 2009;14:536-548. doi:10.1007/s10495-008-0302-x

[8] Zhang X, Liu X, Zhou D, Zheng G. Targeting anti-apoptotic BCL-2 family proteins for cancer treatment. Future Med Chem 2020;12:563-565. doi:10.4155/fmc-2020-0004.

[9] Fischer U, Schulze-Osthoff K. Apoptosis-based therapies and drug targets. Cell Death Differ 2005;12:942-961. doi:10.1038/sj.cdd.4401556

[10] Rathore R, McCallum JE, Varghese E, Florea A-M, Büsselberg D. Overcoming chemotherapy drug resistance by targeting inhibitors of apoptosis proteins (IAPs). Apoptosis 2017;22:898-919. doi:10.1007/s10495-017-1375-1

[11] Augello C, Caruso L, Maggioni M, Donadon M, Montorsi M, Santambrogio R, et al. Inhibitors of apoptosis proteins (IAPs) expression and their prognostic significance in hepatocellular carcinoma. BMC Cancer 2009;9:125. doi:10.1186/1471-2407-9-125

[12] Guttenplan N, Lee C, Frishman WH. Inhibition of Myocardial Apoptosis as a Therapeutic Target in Cardiovascular Disease Prevention. Hear Dis 2001:313-318. doi:10.1097/00132580-200109 000-00007

[13] Rupp H, Maisch B. Control of apoptosis of cardiovascular fibroblasts: A novel drug target. Herz 1999;24:225-231. doi:10.1007/BF03044965

[14] Cai Y, Li Y. Upregulation of miR-29b-3p protects cardiomyocytes from hypoxia-induced apoptosis by targeting TRAF5. Cell Mol Biol Lett 2019;24:27. doi:10.1186/s11658-019-0151-3

[15] Becattini B, Sareth S, Zhai D, Crowell KJ, Leone M, Reed JC, et al. Targeting Apoptosis via Chemical Design. Chem Biol 2004;11:1107-1117. doi:10.1016/j.chembiol.2004.05.022

[16] Wajant H. The TWEAK-Fn14 system as a potential drug target. Br J Pharmacol 2013;170:748-764. doi:10.1111/bph.12337

[17] Lee SM, Yune TY, Kim SJ, Kim YC, Oh YJ, Markelonis GJ, et al. Minocycline

inhibits apoptotic cell death via attenuation of TNF-alpha expression following iNOS/NO induction by lipopolysaccharide in neuron/glia co-cultures. J Neurochem 2004;91:568-578. doi:10.1111/j.1471-4159.2004.02780.x

[18] Tian Y, Wu X, Guo S, Ma L, Huang W, Zhao X. Minocycline attenuates sevoflurane-induced cell injury via activation of Nrf2. Int J Mol Med 2017;39:869-878. doi:10.3892/ijmm.2017.2908

[19] Solesio ME, Peixoto PM, Debure L, Madamba SM, de Leon MJ, Wisniewski T, et al. Carbonic anhydrase inhibition selectively prevents amyloid β neurovascular mitochondrial toxicity. Aging Cell 2018;17:e12787. doi:10.1111/acel.12787

[20] No Title n.d. https://ec.europa.eu/jrc/en/news/2020-cancer-incidence-and-mortality-eu-27-countries.

[21] Schirrmacher V. From chemotherapy to biological therapy: A review of novel concepts to reduce the side effects of systemic cancer treatment (Review). Int J Oncol 2019;54:407-419. doi:10.3892/ijo.2018.4661

[22] Siafaka PI, Üstündağ Okur N, Karavas E, Bikiaris DN. Surface modified multifunctional and stimuli responsive nanoparticles for drug targeting: Current status and uses. Int J Mol Sci 2016;17:1440. doi:10.3390/ijms17091440

[23] Siafaka PI, Üstündağ Okur N, Karantas ID, Okur ME, Gündoğdu EA. Current update on nanoplatforms as therapeutic and diagnostic tools: A review for the materials used as nanotheranostics and imaging modalities. Asian J Pharm Sci 2020. doi:10.1016/j.ajps.2020.03.003

[24] Riganti C, Contino M. New Strategies to Overcome Resistance to Chemotherapy and Immune System in Cancer. Int J Mol Sci 2019;20:4783. doi:10.3390/ijms20194783.

[25] García-Aranda M, Pérez-Ruiz E, Redondo M. Bcl-2 Inhibition to Overcome Resistance to Chemo- and Immunotherapy. Int J Mol Sci 2018;19:3950. doi:10.3390/ijms19123950

[26] Yassemi A, Kashanian S, Zhaleh H. Folic acid receptor-targeted solid lipid nanoparticles to enhance cytotoxicity of letrozole through induction of caspase-3 dependent-apoptosis for breast cancer treatment. Pharm Dev Technol 2020;25:397-407. doi:10.1080/10837450.2019.1703739

[27] Senthil Kumar C, Thangam R, Mary SA, Kannan PR, Arun G, Madhan B. Targeted delivery and apoptosis induction of trans-resveratrol-ferulic acid loaded chitosan coated folic acid conjugate solid lipid nanoparticles in colon cancer cells. Carbohydr Polym 2020;231:115682. doi:10.1016/j.carbpol.2019.115682

[28] Wang H, He L, Zhang P, Zhang J, Chen Z, Ren X, et al. Folate-modified hydroxyapatite nanorods induce apoptosis in MCF-7 cells through a mitochondrial-dependent pathway. New J Chem 2019;43:14728-14738. doi:10.1039/C9NJ03653A

[29] Jin H, Pi J, Yang F, Jiang J, Wang X, Bai H, et al. Folate-Chitosan Nanoparticles Loaded with Ursolic Acid Confer Anti-Breast Cancer Activities in vitro and in vivo. Sci Rep 2016;6:30782. doi:10.1038/srep30782

[30] AbouAitah K, Swiderska-Sroda A, Farghali AA, Wojnarowicz J, Stefanek A, Gierlotka S, et al. Folic acid-conjugated mesoporous silica particles as nanocarriers of natural prodrugs for cancer targeting and antioxidant action. Oncotarget 2018;9:26466-90. doi:10.18632/oncotarget.25470

[31] Murugan C, Rayappan K, Thangam R, Bhanumathi R, Shanthi K, Vivek R, et al. Combinatorial nanocarrier based drug delivery approach for amalgamation of anti-tumor agents in breast cancer cells: an improved nanomedicine strategy. Sci Rep 2016;6:34053. doi:10.1038/srep34053

[32] Chiani M, Norouzian D, Shokrgozar MA, Azadmanesh K, Najmafshar A, Mehrabi MR, et al. Folic acid conjugated nanoliposomes as promising carriers for targeted delivery of bleomycin. Artif Cells, Nanomedicine, Biotechnol 2018;46:757-763. doi:10.1080/21691401.2017.1337029

[33] Mazzotta E, De Benedittis S, Qualtieri A, Muzzalupo R. Actively targeted and redox responsive delivery of anticancer drug by chitosan nanoparticles. Pharmaceutics 2020;12. doi:10.3390/pharmaceutics12010026

[34] Kayani Z, Bordbar A-K, Firuzi O. Novel folic acid-conjugated doxorubicin loaded β-lactoglobulin nanoparticles induce apoptosis in breast cancer cells. Biomed Pharmacother 2018;107:945-956. doi:10.1016/j.biopha.2018.08.047

[35] Schmid D, Fay F, Small DM, Jaworski J, Riley JS, Tegazzini D, et al. Efficient Drug Delivery and Induction of Apoptosis in Colorectal Tumors Using a Death Receptor 5-Targeted Nanomedicine. Mol Ther 2014;22:2083-2092. doi:10.1038/mt.2014.137

[36] Jalilzadeh N, Samadi N, Salehi R, Dehghan G, Iranshahi M, Dadpour MR, et al. Novel nano-vehicle for delivery and efficiency of anticancer auraptene against colon cancer cells. Sci Rep 2020;10:1606. doi:10.1038/s41598-020-58527-0

[37] Wang J, Li N, Cao L, Gao C, Zhang Y, Shuai Q, et al. DOX-loaded peptide dendritic copolymer nanoparticles for combating multidrug resistance by regulating the lysosomal pathway of apoptosis in breast cancer cells. J Mater Chem B 2020;8:1157-1170. doi:10.1039/C9TB02130B

[38] Wang H, Zhang F, Wen H, Shi W, Huang Q, Huang Y, et al. Tumor- and mitochondria-targeted nanoparticles eradicate drug resistant lung cancer through mitochondrial pathway of apoptosis. J Nanobiotechnology 2020;18:8. doi:10.1186/s12951-019-0562-3

[39] Morales-Cruz M, Figueroa CM, González-Robles T, Delgado Y, Molina A, Méndez J, et al. Activation of caspase-dependent apoptosis by intracellular delivery of cytochrome c-based nanoparticles. J Nano biotechnology 2014;12:33. doi:10.1186/s12951-014-0033-9

[40] Yunus U, Zulfiqar MA, Ajmal M, Bhatti MH, Chaudhry G-S, Muhammad TST, et al. Targeted drug delivery systems: synthesis and in vitro bioactivity and apoptosis studies of gemcitabine-carbon dot conjugates. Biomed Mater 2020;15:065004. doi:10.1088/1748-605X/ab95e1

[41] Jiang J-H, Pi J, Jin H, Cai J-Y. Functional graphene oxide as cancer-targeted drug delivery system to selectively induce oesophageal cancer cell apoptosis. Artif Cells, Nanomedicine, Biotechnol 2018;46:S297–S307. doi:10.1080/21691401.2018.1492418

[42] Macfarlane, L., & Murphy, P. R. (2010). MicroRNA: Biogenesis, Function and Role in Cancer. Current Genomics, *11*(7), 537-561. doi:10.2174/138920210793175895

[43] Lima, R. T., Busacca, S., Almeida, G. M., Gaudino, G., Fennell, D. A., & Vasconcelos, M. H. (2011). MicroRNA regulation of core apoptosis pathways in cancer. European Journal of Cancer, *47*(2), 163-174. doi:10.1016/j.ejca.2010.11.005

[44] Lee, Y., Kim, M., Han, J., Yeom, K., Lee, S., Baek, S. H., & Kim, V. N. (2004). MicroRNA genes are transcribed by RNA polymerase II. The EMBO Journal, *23*(20), 4051-4060. doi:10.1038/sj.emboj.7600385

[45] Subramanian, S., & Steer, C. J. (2010). MicroRNAs as gatekeepers of apoptosis. Journal of Cellular Physiology. doi:10.1002/jcp.22066

[46] Friedman, R. C., Farh, K. K., Burge, C. B., & Bartel, D. P. (2008). Most mammalian mRNAs are conserved targets of microRNAs. Genome Research, *19*(1), 92-105. doi:10.1101/gr.082701.108

[47] Kehl, T., Backes, C., Kern, F., Fehlmann, T., Ludwig, N., Meese, E., Lenhof, H. P., & Keller, A. (2017). About miRNAs, miRNA seeds, target genes and target pathways. Oncotarget, *8*(63), 107167-107175. https://doi.org/10.18632/oncotarget.22363

[48] Bartel, D. P. (2009). MicroRNAs: Target Recognition and Regulatory Functions. Cell, *136*(2), 215-233. doi:10.1016/j.cell.2009.01.002

[49] Taghavipour, M., Sadoughi, F., Mirzaei, H., Yousefi, B., Moazzami, B., Chaichian, S., Mansournia, M. A., Asemi, Z. (2020). Apoptotic functions of microRNAs in pathogenesis, diagnosis, and treatment of endometriosis. Cell & Bioscience, *10*(1). doi:10.1186/s13578-020-0381-0

[50] Slattery, M. L., Mullany, L. E., Sakoda, L. C., Wolff, R. K., Samowitz, W. S., & Herrick, J. S. (2018). Dysregulated genes and miRNAs in the apoptosis pathway in colorectal cancer patients. Apoptosis, *23*(3-4), 237-250. doi:10.1007/s10495-018-1451-1

[51] Martinez, N. J., & Walhout, A. J. (2009). The interplay between transcription factors and microRNAs in genome-scale regulatory networks. BioEssays, *31*(4), 435-445. doi:10.1002/bies.200800212

[52] Feng, Z., Zhang, C., Wu, R., & Hu, W. (2011). Tumor suppressor p53 meets microRNAs. Journal of Molecular Cell Biology, *3*(1), 44-50. doi:10.1093/jmcb/mjq040

[53] Singh, R., & Saini, N. (2012). Downregulation of BCL2 by miRNAs augments drug-induced apoptosis - a combined computational and experimental approach. Journal of Cell Science, *125*(6), 1568-1578. doi:10.1242/jcs.095976

[54] Zhou, B., Wang, D., Sun, G., Mei, F., Cui, Y., & Xu, H. (2018). Effect of miR-21 on Apoptosis in Lung Cancer Cell Through Inhibiting the PI3K/ Akt/ NF-κB Signaling Pathway in Vitro and in Vivo. Cellular Physiology and Biochemistry, *46*(3), 999-1008. doi:10.1159/000488831

[55] Zhang, H., Mao, F., Shen, T., Luo, Q., Ding, Z., Qian, L., & Huang, J. (2016). Plasma miR-145, miR-20a, miR-21 and miR-223 as novel biomarkers for screening early-stage non-small cell lung cancer. Oncology Letters, *13*(2), 669-676. doi:10.3892/ol.2016.5462

[56] Egawa, H., Jingushi, K., Hirono, T., Ueda, Y., Kitae, K., Nakata, W., Fujita, K., Uemura, M., Nonomura, N., & Tsujikawa, K. (2016). The miR-130 family promotes cell migration and invasion in bladder cancer through FAK and Akt phosphorylation by regulating PTEN. Scientific reports, *6*, 20574. https://doi.org/10.1038/srep20574

[57] Othman, N., & Nagoor, N. H. (2014). The Role of microRNAs in the Regulation of Apoptosis in Lung Cancer and Its Application in Cancer Treatment. BioMed Research International, *2014*, 1-19. doi:10.1155/2014/318030

Chapter 4

Insights into the Role of Defective Apoptosis in Cancer Pathogenesis and Therapy

Sonia Thapa, Rafiq A. Rather, Shashank K. Singh and Madhulika Bhagat

Abstract

One form of programmed cell death (PCD) is apoptosis. Defective apoptosis is an indispensable causative factor in the development of cancer that allows cancer cells to survive longer and favors the accumulation of oncogenic mutations. Further, upregulation of antiapoptotic proteins (e.g., Bcl-2, Mcl-1) and loss of pro-apoptotic proteins (e.g., Bid, Bad, Bax, Bak) strongly favors apoptosis evasion. The ability of cancer cells to evade apoptosis is critical for the progression and clonal expansion of malignantly transformed cells. Defective apoptosis imparts proliferative advantage to cancer cells or cells with the potential to become cancerous. The mechanisms employed by cancer cells to evade apoptosis can be used in the strategic design of therapeutic regimens aimed at exploiting apoptotic signaling networks to ensure tumor-specific cell death. Therefore, to ensure tumor-specific cell death, we may need to exploit the expression and/or function of different components of apoptotic signaling that are critical for maintaining cell survival and are regulated differently in tumor cells than normal cells. Both inhibitors of anti-apoptotic proteins and activators of pro-apoptotic proteins can be used for cancer therapy. In this chapter, we attempted to summarize the knowledge about the molecular mechanisms of defective apoptosis that could be translated into the development of novel therapeutic agents and therapeutic modalities for cancer treatment.

Keywords: apoptosis, cancer, cell signaling, cancer therapeutics, drugs

1. Introduction

The term apoptosis was coined by Kerr, Wyllie and Currie in 1972 to describe a form of programmed cell death. This type of cell death is mediated by intracellular proteolytic enzyme cascades in a highly regulated fashion. The cells that die as a result of apoptosis typically do not burst and do not release their intracellular contents in the surroundings [1]. As a result apoptosis is typically not associated with inflammation. In multicellular organisms, this process participates in development, tissue homeostasis, and acts as a defense mechanism against the formation of genetically altered cells. Apoptosis is a pivotal for maintaining normal cell turnover and offers defense strategy against tumorigenesis in multicellular organisms. Consequently cells with unrepairable genetic damage or the cells with potential to become cancerous are eliminated via apoptosis. Thus aberrant failure or deficiency of apoptotic signaling can lead to unregulated growth of genetically altered cells and subsequently the development of cancer [2]. Similarly, over activation of apoptosis may result in excessive death of normal cells and may lead to development of neurodegenerative disorders such as autoimmune disorders, Parkinson's disease, Huntington's disease, and Alzheimer's disease [3]. Both failure of apoptosis and excess of apoptosis are detrimental for an organism. Apoptosis is highly programmed biological process that occurs during both physiological and pathological states. An in-depth knowledge of apoptosis is critical for understanding the pathogenesis of many diseases [4]. For example, cancer is a disease condition where little or no apoptosis occurs resulting in unregulated growth of genetically altered cells [5]. Therefore, there is need to identify new drugs and new chemical entities that can potentially target various aspects of apoptosis in tumor cells selectively and specifically. In this chapter, we made an attempt to summarize the knowledge on apoptosis, its molecular mechanism and how defective or unregulated apoptosis leads to cancer development and how apoptosis can be used for therapeutic intervention of cancer. The process of apoptosis manifest multitude of morphological and biochemical changes which can be detected by various cell biological techniques.

1.1 Morphological characteristics of apoptosis

During apoptosis, a cell undergoes a series of morphological changes such as condensation of cytoplasm, condensation of chromatin, nuclear fragmentation, cell rounding, cell shrinkage and blebbing of nuclear and cytoplasmic membranes to form membrane-bound fragments. The chromatin condensation typically begins at the ends of nuclear envelop that forms a ring-like structure [6]. The chromatin condensation continues until it disintegrates into small membrane-bound apoptotic bodies. These small bodies are crowded with closely packed cellular organelles and smaller fragments of nucleus [6, 7]. These apoptotic bodies are immediately taken up by phagocytes such as macrophages, dendritic cells and Langerhans cells. The engulfment of these apoptotic bodies by phagocytes occurs without any inflammation and release of intracellular contents. As we know, apoptosis is usually considered a non-inflammatory process while necrosis (another mode of cell death) activates inflammation. This apoptotic process is typically mediated by the proteolytic cleavage of cellular substrates by caspases, and signaling elements [7, 8]. The morphological changes occur in parallel with the activation of a number of complex biochemical effector pathways that cause solubilization of the apoptotic cells.

1.2 Biochemical characteristics of apoptosis

Besides morphological changes, an apoptotic cell progresses through a series of biochemical changes such as endonucleolytic fragmentation of genomic DNA by endogenous DNases which cleave intranucleosomal regions of genomic DNA into double stranded DNA (dsDNA) fragments of sizes varying from 180 bp to 200 bp. DNA fragmentation is the frequent end point and the widespread marker of apoptosis [9]. The DNA fragments that are formed during apoptosis contain single base 3′ overhangs as well as blunt ends. The DNA fragments formed during apoptosis are noticeable as a ladder pattern in the ethidium based-electrophoresis of genomic DNA [10]. The most important enzymes that catalyze the cleavage of genomic DNA during apoptosis include DNA fragmentation factor (DFF40), caspase activated DNase (CAD) and 70-kDa endonuclease (NUC70). In a normal healthy cell, DFF40 and CAD are retained as inactive heterodimers with inhibitor proteins DFF45 and ICAD (inhibitor of CAD). During apoptosis, these DNase are selectively activated upon cleavage by caspase 3. Once activated, DFF40 and CAD are sufficient to induce the nuclear morphological changes characteristic of apoptosis. Similarly, when isolated HeLa nuclei are incubated with NUC70, the nuclei undergo inter-nucleosomal DNA fragmentation. NUC70 is a cytoplasmic endonuclease that is translocated to the nucleus after the initiation of apoptosis [7, 11].

2. Pathways of apoptosis

There are atleast two fundamental pathways of apoptosis: extrinsic, death receptor-mediated pathway and intrinsic, mitochondria-mediated pathway. Another important pathway of apoptosis is perforin/granzyme pathway. Apoptosis can be initiated through any one of these pathways.

2.1 The extrinsic death receptor pathway

This pathway begins when ligand binds to the death receptor, for example, with the attachment of extracellular ligands, like tumor necrosis factor (TNF), Fas ligand (Fas-L), TNF-related apoptosis-inducing ligand (TRAIL) to the extracellular domain of transmembrane receptors. There are six death receptors (TNFR1, Fas, DR3, DR4 [TRAILR1], DR5 [TRAILR2], and DR6) that have been identified in mammalian cells (**Table 1**). While signaling through TNFR1 and DR3 is proinflammatory in nature, signaling via other death receptors is predominantly pro-apoptotic. The death receptors bind via their intracellular death domain with adapter proteins such as Fas-associated death domain (FADD) and TNF receptor-associated death domain (TRADD). These adaptor proteins, called the Death effector Domain (DED) have another protein interaction domain [12]. These sequential steps leads to the formation of a death inducing signaling complex (DISC), that leads to auto-catalytic activation of procaspase-8 (**Figure 1**). Active caspase-8 activates effector/executioner caspases (*c*ysteine-dependent *as*partyl-specific prote*ases*), which cause cell death by damaging the nucleus and other intracellular structures. Once caspase-8 is activated, the execution phase of apoptosis is triggered and activated. This type of apoptosis, which is death receptor-mediated apoptosis, can be inhibited by a protein called c-FLIP. c-FLIP bind to FADD and caspase-8, and turns them ineffective. A protein called Toso, has been shown to block Fas-induced apoptosis in T cells. This may be due to inhibition of caspase-8. The death receptors belong to the tumor necrosis factor (TNF) receptor gene superfamily. TNF family contains cysteine-rich extracellular domains and a cytoplasmic domain of about 80 amino acids called the “death domain”. This death domain plays a critical role in transmitting the death signal from the cell surface to the intracellular signaling pathways [13–15].

Death receptor	Activating ligand
TNFR1/DR1/CD120a/p55	TNF
Fas/CD95/Apo1/DR2	Fas/CD95/Apo1/DR2
DR3/Apo3/WSL-1/TRAMP/LARD	Apo3L/TWEAK
TRAIL-R1/DR4	TRAIL/Apo2L
DR6	TRADD

Table 1.
Death receptors and their cognate ligands.

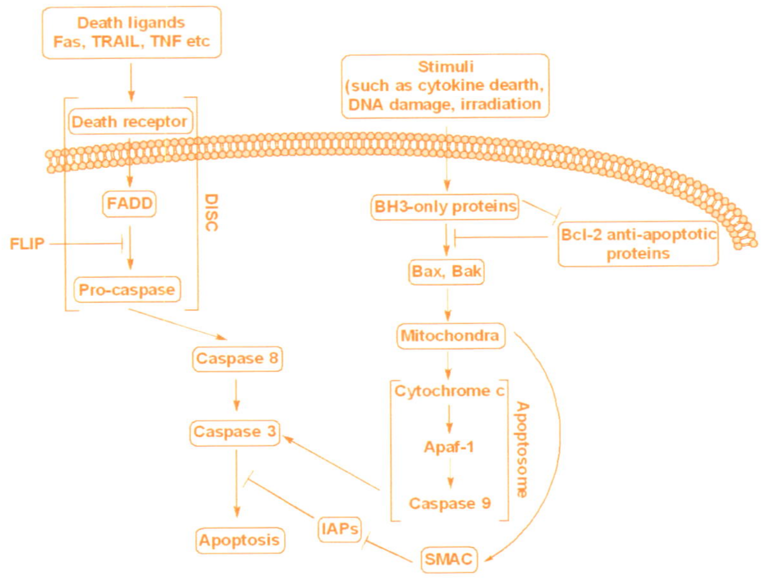

Figure 1.
Key components of the intrinsic and extrinsic pathways of apoptosis.

2.2 The intrinsic mitochondrial apoptosis pathway

Intrinsic apoptosis is a form of regulated cell death that is activated in response to a variety of stimulus such as dearth of growth factors, DNA damage, endoplasmic reticulum (ER) stress, and overproduction of reactive oxygen species (ROS). The intrinsic pathway is affected by members of the Bcl-2 family proteins such as Bcl-2 and Bax. These proteins usually are bound to the outer mitochondrial membrane and act as antiapoptotic and pro-apoptotic proteins respectively [6]. In the intrinsic pathway the main consequence of proapoptotic signaling is mitochondrial membrane perturbation and release of cytochrome 'c' in the cytoplasm. Once released in the cytosol, cytochrome 'c' forms a complex with apoptotic protease activating factor 1 (APAF1) and inactive form of pro-caspase 9, commonly known as apoptosome (**Figure 1**). This complex hydrolyzes adenosine triphosphate (ATP) to cleave and activate caspase 9. The initiator caspase 9 then cleaves and activates the executioner caspase 3, caspase 6, and caspase 7, resulting in cell apoptosis. The antiapoptotic proteins Bcl-2 and Bcl-X_L inhibit cytochrome 'c' release [8]. The Bcl-2 group of proteins share one to four Bcl2 homology (BH) domains (i.e., BH1, BH2, BH3, and BH4) [16]. Bax and Bak of the Bcl-2 family form pores across the outer mitochondrial membrane. Besides this, Bax also continuously cycles between the outer mitochondrial membrane (OMM) and the cytosol, and exhibits a quiescent inactive dimeric conformation [17]. Bak constitutively resides at the OMM, where it inserts within hydrophobic C-terminal of voltage dependent anion channel 2 (VDAC2). Upon apoptosis induction, mitochondrial pools of Bax and Bak undergo direct or indirect activation by pro-apoptotic proteins [16].

The BH3-only proteins are BCL2 binding component 3 or p53, Bcl-2-like 11 (BIM) [18]. Some internal stimuli cause an increase in mitochondrial membrane permeability and these stimulus send the signal to the mitochondria, that form mitochondrial outer membrane permeabilization (MOMP). MOMP is most commonly mediated via a variety of protein membrane and protein–protein interactions of the B-cell lymphoma 2 protein (Bcl-2) family [19–21].

Both extrinsic and intrinsic pathways terminate at the execution phase in the final stage of apoptosis. Activated execution caspases further activates cytoplasmic endonucleases, proteases that degrade nuclear and cytoplasmic proteins respectively. The executioner caspases are caspase-3, caspase-6, and caspase-7 that cleave substrates like cytokeratins, PARP, and others that results in morphological and biochemical changes in apoptotic cells. The most important of the executioner caspases is caspase-3 and is activated by any of the initiator caspases (caspase-8, caspase-9, or caspase-10). Caspase-3 induces cytoskeleton reorganization and disintegration of the cell into apoptotic bodies [22].

2.3 Perforin/granzyme pathway

The granzymes belong to a family of serine proteases contained in cytotoxic granules of innate and adaptive immune killer cells. The key function of these enzymes is eliminating viruses and tumor cells. They also regulate immune cells and inflammation by controlling the survival of lymphocytes. Granzymes are expressed by three gene clusters. Granzyme A and granzyme B are the most abundant granzymes. Granzyme-mediated apoptosis is mainly used by lymphocytes to destroy virus infected cells. It is a kind of type IV hypersensitivity where sensitized CD8+ cells kill antigen-bearing cells. The FasL/FasR interaction is the predominant method of CTL-induced apoptosis. They also involve pathways that include secretion of the transmembrane pore-forming molecule perforin with a subsequent release of cytoplasmic granules that have serine proteases as well as granzyme B, through the pore and into the target cell [23]. Granzyme B cleaves proteins at aspartate residues and can activate pro-caspase-10. It can also cleave factors like ICAD (Inhibitor of Caspase Activated DNAse). Granzyme B can amplify the death signal by inducing cytochrome c release. It can also activate caspase-3 directly [24].

Therefore, in these types of pathways, signaling pathways that are upstream are bypassed and execution phase of apoptosis is directly influenced. It is suggested that both the mitochondrial pathway and direct activation of caspase-3 are critical for granzyme B induced killing activates both mitochondrial pathway and caspase-3 activation [24, 25]. The immune cells that express highly variable and regulated patterns of granzymes include natural killer (NK) cells, cytotoxic CD4 and CD8 T cells, and regulatory T cells (Tregs). The granules contain perforin, to deliver the granzymes into the target cell. When cytotoxic T lymphocytes (CTLs) and NK cells form an immune synapse with a target cell for its elimination, cytotoxic granules join immune synapse where its membrane fuses with the killer cell membrane. This action results in the release of the granule contents into the synaptic cleft. The granzymes are then initiate distinct pathways of programmed cell death [8, 26, 27].

3. Bio markers of apoptosis

A biomarker is measured and evaluated to indicate normal or diseased biological processes. It has the potential to enhance translational progress and accelerate drug development. They allow monitoring of drug efficacy and also help in preclinical drug evaluation. It also allows early detection of toxicity during drug evaluation. The fragmentated DNA on agarose gel is a usual marker to detect apoptosis. The poly (ADP- ribose) polymerase (PARP) cleaved form, observed in cells undergoing apoptosis. Other markers are cytokeratin-18 cleavage by caspase 3. It may be detected by using the antibody M30. Cleavage of various caspases can be detected using flow cytometry during apoptosis [28, 29]. The widely used biomarkers of apoptosis, their methods of analysis and the specimen needed for analysis are discussed in **Table 2**. However, the use of these biomarkers as a tool to predict the occurrence of apoptosis in the pathogenesis of different diseases states warrants further investigation.

Biomarkers	Analysis method	Specimen for biomarker analysis
Cytochrome c	ELISA	Biomarker in blood samples
Activated caspases Caspase 2,3,7,8,9	ELISA Flow cytometry	Detection by substrate active site activation
Cytokeratins	ELISA Flow cytometry	Biomarker in blood samples
Bcl-2, Bcl-x_L	ELISA Flow cytometry	Its over expression results in chemoresistance
Externalized Phosphatidylserine	ELISA Flow cytometry	Early apoptotic event. It measures annexin that binds to the external ligand
Fas, Fas ligand	ELISA Flow cytometry	Expression on B and T cells, also expressed on normal and tumor tissues.
Granzyme B	PET imaging	Response in immunotherapy
8-hydroxy-2-deoxyguanosine, 3-nitrotyrosine	ELISA	Biomarker in tissue or urine

Table 2.
Biomarkers of apoptosis.

4. Role of apoptosis in cancer

Cells can die in variety of ways which includes apoptosis, necrosis, mitotic catastrophe, senescence, and autophagy. Of these different modes of cell death, apoptosis is active, programmed and genetically controlled. While physiological apoptosis helps to eliminate genetically altered cells, defective apoptosis is intimately connected with cancer pathogenesis [30]. Apoptotic signaling regulation is important to preserve a proper balance between cell death and cell survival. It is also important in maintaining genome integrity. The Apoptosis evasion is a prominent hallmark of cancer and cancer cells can use a number of diverse strategies to evade apoptosis. The disruption in the balance between pro- and anti-apoptotic proteins contributes to carcinogenesis [31]. It may be due to the reducing apoptosis in malignant cells [30]. For example, the imbalance between pro- and anti-apoptotic Bcl-2 proteins, its genetic and epigenetic alterations, can promote cancer cell survival. The elevated levels of anti-apoptotic family members is a distinct mechanism of apoptosis dysregulation in cancer. The anti-apoptotic proteins are widely over-expressed in cancer cells to overcome stress signals. Over-expression of anti-apoptotic Bcl-2 family proteins is often correlated with recurrence, poor prognosis, and resistance to cancer therapeutics [11].

miRNAs are a class of non-coding RNAs that regulate post-transcriptional gene expressions and silence target mRNAs. miRNAs dysregulation are associated with different human cancers and microRNAs (miRNAs) can function as oncogenes as well as tumor suppressors and their dysregulation are associated with many different human cancers.. miRNAs target different mRNAs and act as anti-apoptotic or pro-apoptotic regulatots that involved in the apoptotic pathways [32]. For example, miR-15/16 targets the anti apoptotic factor (Bcl-2). In many cancers, the mostly up-regulated miRNA is anti-apoptotic miRNA-21. It targets the programmed cell death 4 gene (PCD 4 gene or PDCD4), tropomyosin 1 (TPM1), and the phosphatase and tensin homolog (PTEN), to modulate apoptosis. The application of anti-microRNAs imitate may act as a potent therapeutic strategy to inhibit key molecular signaling pathways that are present in cancer [32, 33].

Cancer cells often show dysregulated expression patterns of diverse long non-coding RNAs (lncRNAs) in specific kinds of tumors. Therefore, their upexpression or down-regulation in cancer cells often sensitizes cells to apoptotic treatments.

They also induce and modulate apoptosis. Therefore, targeting lncRNAs in cancer cells can be utilized for cancer treatment. Cellular-FLICE inhibitory protein (c-FLIP) is a critical negative regulator of the apoptotic pathway. The apoptosis regulation due to c-FLIP, mediated by the death receptors Fas, TNF-R1, DR4 and DR5. It exists in three isoforms that are derived from diverse mRNA splice variants. These splice variants are transcribed under the same promoter, namely c-FLIPL, c-FLIPS, and c-FLIPR. The three isoforms of c-FLIP acts at the DISC level and inhibit the procaspase 8 and 10 activation. C-FLIP high expression is found in many cancers and its downexpression can restore apoptosis mediated by TRAIL Aand CD95L. Thus, c-FLIP can act as a promising target for cancer therapy [34]. c-FLIP can also induce apoptosis at low and more physiologically relevant expression levels by recruiting at the DISC to increase caspase-8 activation.

NF-kB activation by CD40 ligand or TNF-α results in overexpression of c-FLIP and the prohibition of TNFR1, Fas- and TRAIL receptors induces apoptosis. Activation of several pathways, such as mitogen-activated protein kinase (MAPK), the phosphatidylinositol-3 kinase (PI3K)/Akt, can enhance the expression of c-FLIP and can hamper apoptosis induced by death receptors [35].

IAPs increase cell survival during cellular stresses such as ER stress and prevent both intrinsic and extrinsic apoptosis. Caspases dysregulation may inhibit apoptosis and carcinogenesis. Down-regulation of different caspases has been observed in many cancers. For example, caspase-9 and caspase-3 downregulation can leads to formation of different forms of cancers such as colorectal, ovarian, breast, and cervical cancers [36, 37].

5. Therapeutic targeting of defective apoptosis in cancer

Apoptosis plays a pivotal role in cancer pathogenesis. Therefore, understanding the molecular basis of defective apoptosis has garnered tremendous attention from medical researchers. Novel therapeutic agents and treatment modalities to modulate disease pathogenesis are under trials in both preclinical and clinical settings. An important strategy to modulate cancer pathogenesis involves the use of small molecule inhibitors (SMIs) that target specific components in apoptotic cascades.

5.1 Targeting anti-apoptotic Bcl-2 family members

Bcl-2 is over expressed in many types of tumors and imparts therapeutic intractability to such tumors. A highly selective inhibitor of Bcl-2 is Venetoclax. This drug has been approved for routine clinical practice and is currently in use for the treatment of ALL (acute lymphocytic leukemia), CLL (chronic lymphocytic leukemia), and multiple myeloma T-cell prolymphocytic leukemia. Venetoclax is a Bcl2-selective BH3-mimetic. BH3-mimetics represents a class of anticancer drug that mimic the functions of BH3-only proteins. These mimetics bind to prosurvival proteins like Bcl-2 and inhibit their ability to bind Bax or Bak [38]. Thus, when Bcl-2 overexpressing cancer cells are treated with venetoclax *in vitro*, the cancer cells undergo apoptosis. Venetoclax is often used alone or in combination with other drugs such as rituximab, ibrutinib, azacitidine/decitabine, and bortezomib/dexamethasone against various hematological malignancies. Bcl-2 family members over expression are affiliated with aggressive cancer and chemo resistive [39]. These credentials make these proteins as highly encouraging therapeutic targets to develop pharmacological anticancer drugs. Bcl-2 family members inhibition by small interfering RNAs (siRNAs) may also induce apoptosis and can reduce tumor growth [40]. For example, Mcl-1 down regulation by siRNA induced significant apoptosis in leukemia cells. Many different microRNAs have been associated that regulate Bcl-2 expression such as miR-195, miR-24-2, and miR-365-2 act as negative regulators of Bcl-2 family, which shows the therapeutic potential of these miRNAs [16, 41].

5.2 Mcl-1 inhibitors

The anti-apoptotic protein myeloid cell leukemia-1 (Mcl-1) is an important regulator of apoptosis and a central driver of drug resistance in multitude of human malignancies. Overexpressed Mcl-1 imparts drug resistance of both solid tumors and hematological malignancies against various therapeutic agents. Several inhibitors have been developed that show promising anticancer activities in preclinical and clinical settings. For example S63845 induces apoptosis in SCLC cell lines in vitro at an IC50 of 23 to 78 nM, while in xenograft models this molecule causes significant reduction in tumor volumes. S63845 can be used in combination with navitoclax (a dual inhibitor of Bcl-x_L and Bcl-2) where S63845 reduced the cell viability of SCLC cells and showed synergistic effects in S63845-resistant xenograft models [42]. Furthermore, Mcl-1 is an attractive drug target in lung cancer due to its non-apoptotic involvement in DNA repair. Thus targeting Mcl-1 with a small molecule inhibitor (MI-233) blocks Mcl-1-mediated HR DNA repair and thereby sensitizes cancer cells to treatment induced replication stress. MI-233 shows strong synergism with hydroxyurea or olaparib in lung cancer models [43]. Similarly, targeted Mcl-1 inhibition by RNAi also increases caspase-mediated cell death in cell models such as ERα + breast cancer cells [44]. A specific Mcl-1 inhibitor VU661013 induce tumor cell death and a causes synergistic reduction when used in combination with ABT-263 in tumor volume [45]. During the last several years, many Mcl-1 inhibitors have been developed. However, due to large surface-exposed hydrophobic BH3 binding groove, specific targeting of Mcl-1 poses a big challenge. Thus indirect targeting of Mcl-1 is emerging as important mechanism of action of alternative drug classes such as CDK9 inhibitors or deubiquitination inhibitors [46]. The pharmacological characteristics of the major Mcl-1 inhibitors are tabulated in **Table 3**.

5.3 X-linked inhibitor of apoptosis protein (XIAP) inhibitors

Although IAPs control plethora of signaling pathways, they were initially thought to be responsible for caspase inhibition by acting as negative regulators of apoptosis [36, 37]. Many small molecules have been designed to target the IAP proteins and one such category of an molecules is Smac mimetics (SMs). A recently developed XIAP inhibitor BMT-062789 displays remarkable anticancer activity against a panel of lymphoma cell lines. BMT-062789 is a heterodimeric mimetic of the second mitochondrial activator of caspases (SMAC) [47]. This molecule inhibits both the caspase 9 and caspase 3/7 binding domains of XIAP [48]. When used in combination with etoposide, BMT-062789 induces apoptosis in rituximab resistant cell line models Raji 4RH and RL 4RH. ASTX660 is another orally bioavailable, non-peptidomimetic antagonist of both XIAP and cellular IAP1 (cIAP1), with potential antineoplastic and pro-apoptotic activities. ASTX660 selectively binds to and inhibits the activity of XIAP and cIAP1 [49].

5.4 Caspase activators in cancer therapy

Caspases constitutes a group of cytosolic aspartate-specific cysteine proteases that participate in the initiation and execution of apoptosis Pharmacological activation of caspases using small molecule activators is an effective therapeutic

Drug	Affinity (Ki)	Activity in tumors
S63845	Ki < 1.2 nM	non-small cell lung cancer, breast cancer, melanoma
AMG 176	Ki = 0.06 nM	breast cancer
AZD5991	Ki = 0.2 nM	multiple myeloma, acute myeloid leukemia
VU661013	Ki = 97 ± 30 pM	acute myeloid leukemia
Compound 42	Ki = 0.03 nM	triple-negative breast cancer, hematological malignancies
β-carboline copper(II) complexes	Ki = 1.2–96.4 nM	non-small cell lung cancer

Table 3.
Pharmacological characteristics of some Mcl-1 inhibitors.

Caspase activator	Mechanism of action
Cisplatin	potent pro-apoptotic anticancer agent; activates caspase-3
α-(Trichloromethyl)-4-pyridineethanol (PETCM)	activator of caspase-3
Apoptosis activator 2	promotes apoptosome formation and activates caspase-9/caspase-3 pathway; selectively induces tumor cell apoptosis
4-(Phenylmethyl)-1-piperazineacetic acid [[2-hydroxy-3-(2-propenyl)phenyl]methylene] hydrazide (PAC 1)	pro-apoptotic; activator of procaspase-3
Gambogic acid	apoptosis inducer; activates caspases and inhibits Bcl-2 family proteins

Table 4.
Caspase activators.

strategy to kill cancer cells and can potentially help to reverse drug resistance. The most important cysteine protease in the caspase cascade is caspase 3 [50]. In a normal cell, capase-3 is typically in anactive form by an intramolecular electrostatic interaction favored by triplet of aspartic acid residues (also known as safety-catch). Although, there are indication that cell death can proceed in the absence of caspases, activation caspase family members is critical for execution of cell death in apoptotic cells. Several drugs for activating caspases exist (**Table 4**). An important caspase-inducing agent is apoptin, which is a proline-rich protein capable of inducing apoptosis in cancer cells in a selective manner. This protein is obtained from chicken anemia virus that causes tumor-specific apoptosis without interfering with normal cells. As such, apoptin is considered as a highly tumor-specific therapeutic agent [51, 52]. This drug is still in preclinical testing.

6. Conclusion

Due to its role in tissue homeostasis and cancer pathogenesis, apoptosis has received tremendous attention as a target for therapeutic intervention of cancer. Thus targeting apoptotic signaling networks is an important therapeutic strategy for cancer therapy. Many drugs have been developed that modulate different components of apoptotic signaling. However, many tumors ultimately develop drug resistance against these drugs. Therefore there is a need to develop novel therapeutic agents and new treatment strategies that can reverse the drug-resistant phenotype of tumors and render these tumors sensitive towards therapeutic regimens. A limited number of FDA-approved drugs exist that directly target apoptotic pathways. Many of these drugs target the Bcl-2 family member such as Bcl-2 itself and Mcl-1. Caspase 3 activators are also being tested in preclinical settings. Other therapeutic strategies involve the use of drugs that activate the extrinsic pathway of apoptosis or tumor suppressor pathways or modulate tumor microenvironment. However, it is interesting to conclude that apoptosis can be induced in cancer cells by activating both the intrinsic and extrinsic pathways of apoptosis. However there is a need to identify therapeutic targets that are unique to cancer cells so that their modulation may not affect the normal surrounding cells. Thus, while targeting apoptotic signaling networks, there is a need to ensure tumor-specific apoptosis without inducing apoptosis in normal cells. We are hopeful that, in future, a better understanding of apoptotic signaling may lead to the development of novel anticancer drugs that could target different components of apoptotic signaling selectively and specifically. However, the lack of precise relationship between level of proteins involved in apoptosis and the clinical outcome poses a big challenge in the development of novel therapeutic agents and as such warrants in-depth understanding of apoptotic signaling networks.

Acknowledgements

The first author is thankful to University Grants Commission, New Delhi, India for junior research fellowship (UGC-JRF). The corresponding author is grateful to SERB-National Postdoctoral Fellowship program and *Science and Engineering Research Board, Department of Science and Technology*, New Delhi, India (Grant Number: PDF/2016/002730) for financial support.

Conflict of interest statement

The authors have declared that there are no conflicts of interest.

Author contributions

RAR and ST conceived the idea. ST wrote the chapter with the assistance from RAR, SKK and MB. All authors contributed to the chapter and approved the submitted version.

Author details

Sonia Thapa[1,2], Rafiq A. Rather[3]*, Shashank K. Singh[1,2] and Madhulika Bhagat[3]

1 Cancer Pharmacology Division, CSIR-Indian Institute of Integrative Medicine, Jammu, India

2 Academy of Scientific and Innovative Research (AcSIR), Ghaziabad, India

3 School of Biotechnology, University of Jammu, Jammu and Kashmir, India

*Address all correspondence to: rafiqueahmadd@gmail.com

References

[1] Wyllie, A.H., *Apoptosis: an overview*. British medical bulletin, 1997. **53**(3): p. 451-465.

[2] Lowe, S.W. and A.W. Lin, *Apoptosis in cancer*. Carcinogenesis, 2000. **21**(3): p. 485-495.

[3] Mattson, M.P., *Apoptosis in neurodegenerative disorders*. Nature reviews Molecular cell biology, 2000. **1**(2): p. 120-130.

[4] Mattson, M.P., *Neuronal life-and-death signaling, apoptosis, and neurodegenerative disorders*. Antioxidants & redox signaling, 2006. **8**(11-12): p. 1997-2006.

[5] Fernald, K. and M. Kurokawa, *Evading apoptosis in cancer*. Trends in cell biology, 2013. **23**(12): p. 620-633.

[6] D'Arcy, M.S., *Cell death: a review of the major forms of apoptosis, necrosis and autophagy*. Cell biology international, 2019. **43**(6): p. 582-592.

[7] Majtnerová, P. and T. Roušar, *An overview of apoptosis assays detecting DNA fragmentation*. Molecular biology reports, 2018. **45**(5): p. 1469-1478.

[8] Jan, R., *Understanding apoptosis and apoptotic pathways targeted cancer therapeutics*. Advanced pharmaceutical bulletin, 2019. **9**(2): p. 205.

[9] Wyllie, A., G. Beattie, and A. Hargreaves, *Chromatin changes in apoptosis*. The Histochemical Journal, 1981. **13**(4): p. 681-692.

[10] Allen, R.T., W.J. Hunter III, and D.K. Agrawal, *Morphological and biochemical characterization and analysis of apoptosis*. Journal of pharmacological and toxicological methods, 1997. **37**(4): p. 215-228.

[11] Chen, L., Y. Zeng, and S.-F. Zhou, *Role of apoptosis in cancer resistance to chemotherapy*. Current Understanding of Apoptosis-Programmed Cell Death, 2018.

[12] Liu, B., et al., *Both intrinsic and extrinsic apoptotic pathways are involved in Toll-like receptor 4 (TLR4)-induced cell death in monocytic THP-1 cells*. Immunobiology, 2017. **222**(2): p. 198-205.

[13] Mandal, P., A.L. McCormick, and E.S. Mocarski, *TNF Signaling Dictates Myeloid and Non-Myeloid Cell Crosstalk to Execute MCMV-Induced Extrinsic Apoptosis*. Viruses, 2020. **12**(11): p. 1221.

[14] Salva, K.A., et al., *Epigenetically Enhanced PDT Induces Significantly Higher Levels of Multiple Extrinsic Pathway Apoptotic Factors than Standard PDT, Resulting in Greater Extrinsic and Overall Apoptosis of Cutaneous T-cell Lymphoma*. Photochemistry and photobiology, 2018. **94**(5): p. 1058-1065.

[15] Gonçalves, I., et al., *sTRAIL-R2 (soluble TNF [tumor necrosis factor]-related apoptosis-inducing ligand receptor 2) a marker of plaque cell apoptosis and cardiovascular events*. Stroke, 2019. **50**(8): p. 1989-1996.

[16] Campbell, K.J. and S.W. Tait, *Targeting BCL-2 regulated apoptosis in cancer*. Open biology, 2018. **8**(5): p. 180002.

[17] Inoue-Yamauchi, A., et al., *Targeting the differential addiction to anti-apoptotic BCL-2 family for cancer therapy*. Nature communications, 2017. **8**(1): p. 1-14.

[18] Chen, W.H., G.F. Luo, and X.Z. Zhang, *Recent advances in subcellular targeted cancer therapy based on functional materials*. Advanced Materials, 2019. **31**(3): p. 1802725.

[19] Kalkavan, H. and D.R. Green, *MOMP, cell suicide as a BCL-2 family*

business. Cell Death & Differentiation, 2018. **25**(1): p. 46-55.

[20] Kale, J., E.J. Osterlund, and D.W. Andrews, *BCL-2 family proteins: changing partners in the dance towards death.* Cell Death & Differentiation, 2018. **25**(1): p. 65-80.

[21] Pihán, P., A. Carreras-Sureda, and C. Hetz, *BCL-2 family: integrating stress responses at the ER to control cell demise.* Cell Death & Differentiation, 2017. **24**(9): p. 1478-1487.

[22] Heckmann, B.L., B. Tummers, and D.R. Green, *Crashing the computer: apoptosis vs. necroptosis in neuroinflammation.* Cell Death & Differentiation, 2019. **26**(1): p. 41-52.

[23] Zhou, Z., et al., *Granzyme A from cytotoxic lymphocytes cleaves GSDMB to trigger pyroptosis in target cells.* Science, 2020. **368**(6494).

[24] Velotti, F., et al., *Granzyme B in Inflammatory Diseases: Apoptosis, Inflammation, Extracellular Matrix Remodeling, Epithelial-to-Mesenchymal Transition and Fibrosis.* Frontiers in immunology, 2020. **11**: p. 2828.

[25] Chiusolo, V., et al., *Granzyme B enters the mitochondria in a Sam50-, Tim22-and mtHsp70-dependent manner to induce apoptosis.* Cell Death & Differentiation, 2017. **24**(4): p. 747-758.

[26] Wu, M., et al., *Phosphorylation of SET mediates apoptosis via P53 hyperactivation and NM23-H1 nuclear import.* Neurobiology of aging, 2018. **69**: p. 38-47.

[27] Mátyási, B., et al., *The function of NM23-H1/NME1 and its homologs in major processes linked to metastasis.* Pathology & Oncology Research, 2020. **26**(1): p. 49-61.

[28] Nosek, H., et al., *Tumor Necrosis Factor-Like Weak Inducer of Apoptosis and Selected Cytokines—Potential Biomarkers in Children with Solitary Functioning Kidney.* Journal of Clinical Medicine, 2021. **10**(3): p. 497.

[29] Shermatov, K., et al., *Levels of Serum M30 and M65 Proteins as Biomarkers of Apoptosis in Children Exposed To Passive Smoking.* Konuralp Tıp Dergisi, 2018. **10**(3): p. 289-293.

[30] Burke, P.J., *Mitochondria, bioenergetics and apoptosis in cancer.* Trends in cancer, 2017. **3**(12): p. 857-870.

[31] Manchanda, A. and A. Arora, *Role of Apoptosis in Cancer Proliferation and in molecular target therapy.* Baba Farid University Dental Journal, 2019. **9**(2): p. 67-72.

[32] Shirjang, S., et al., *MicroRNAs in cancer cell death pathways: Apoptosis and necroptosis.* Free Radical Biology and Medicine, 2019. **139**: p. 1-15.

[33] Slattery, M.L., et al., *Dysregulated genes and miRNAs in the apoptosis pathway in colorectal cancer patients.* Apoptosis, 2018. **23**(3): p. 237-250.

[34] Zhang, G., et al., *Identification of cancer-related miRNA-lncRNA biomarkers using a basic miRNA-lncRNA network.* PLoS One, 2018. **13**(5): p. e0196681.

[35] Zhang, W., et al., *MiR-126 reverses drug resistance to TRAIL through inhibiting the expression of c-FLIP in cervical cancer.* Gene, 2017. **627**: p. 420-427.

[36] Rathore, R., et al., *Overcoming chemotherapy drug resistance by targeting inhibitors of apoptosis proteins (IAPs).* Apoptosis, 2017. **22**(7): p. 898-919.

[37] Mohamed, M.S., et al., *Inhibitors of apoptosis: clinical implications in cancer.* Apoptosis, 2017. **22**(12): p. 1487-1509.

[38] Trisciuoglio, D. and D. Del Bufalo, *New insights into the roles of antiapoptotic members of the Bcl-2 family in melanoma progression and therapy*. Drug discovery today, 2021.

[39] Suvarna, V., V. Singh, and M. Murahari, *Current overview on the clinical update of Bcl-2 anti-apoptotic inhibitors for cancer therapy*. European journal of pharmacology, 2019. **862**: p. 172655.

[40] Kim, E.M., et al., *The p53/p21 complex regulates cancer cell invasion and apoptosis by targeting Bcl-2 family proteins*. Cancer research, 2017. **77**(11): p. 3092-3100.

[41] Garner, T.P., et al., *Progress in targeting the BCL-2 family of proteins*. Current Opinion in chemical biology, 2017. **39**: p. 133-142.

[42] Mattox, T.E., et al., *Novel RAS inhibitor, MCI-062, potently and selectively inhibits the growth of KRAS mutant pancreatic tumor cells by blocking GTP loading of RAS*. Cancer Res., 2019.

[43] Cidado, J., et al., *AZD4573 is a highly selective CDK9 inhibitor that suppresses MCL-1 and induces apoptosis in hematologic cancer cells*. Clinical cancer research, 2020. **26**(4): p. 922-934.

[44] Stewart, M.L., et al., *The MCL-1 BH3 helix is an exclusive MCL-1 inhibitor and apoptosis sensitizer*. Nature chemical biology, 2010. **6**(8): p. 595-601.

[45] Luedtke, D.A., et al., *Inhibition of Mcl-1 enhances cell death induced by the Bcl-2-selective inhibitor ABT-199 in acute myeloid leukemia cells*. Signal transduction and targeted therapy, 2017. **2**(1): p. 1-9.

[46] Hird, A.W. and A.E. Tron, *Recent advances in the development of Mcl-1 inhibitors for cancer therapy*. Pharmacology & therapeutics, 2019. **198**: p. 59-67.

[47] Dubrez-Daloz, L., A. Dupoux, and J. Cartier, *IAPs: more than just inhibitors of apoptosis proteins*. Cell cycle, 2008. **7**(8): p. 1036-1046.

[48] Holcik, M., H. Gibson, and R.G. Korneluk, *XIAP: apoptotic brake and promising therapeutic target*. Apoptosis, 2001. **6**(4): p. 253-261.

[49] Schimmer, A., et al., *Targeting XIAP for the treatment of malignancy*. Cell Death & Differentiation, 2006. **13**(2): p. 179-188.

[50] Fulda, S. and K.-M. Debatin, *Caspase activation in cancer therapy*, in *Madame Curie Bioscience Database [Internet]*. 2013, Landes Bioscience.

[51] Maddika, S., et al., *Cancer-selective therapy of the future: apoptin and its mechanism of action*. Cancer biology & therapy, 2006. **5**(1): p. 10-19.

[52] Los, M., et al., *Apoptin, a tumor-selective killer*. Biochimica et Biophysica Acta (BBA)-Molecular Cell Research, 2009. **1793**(8): p. 1335-1342.

Chapter 5

Efferocytosis: An Interface between Apoptosis and Pathophysiology

Bichandarkoil Jayaram Pratima and Namasivayam Nalini

Abstract

Several cell death modes, each with a unique feature and mode of inducing cell death have been established. Cell death occurring under physiological conditions is primarily caused by apoptosis, which is a non-inflammatory or silent process, whereas necroptosis or pyroptosis is triggered by pathogen invasion, which stimulates the immune system and induces inflammation. In physiology, clearing dead cells and associated cellular debris is necessary since billions of cells die during mammalian embryogenesis and every day in adult organisms. For degradation, dead cells produced by apoptosis are quickly engulfed by macrophages. This chapter will present a description of the phagocytosis of dead and dying cells, by a process known as efferocytosis. Macrophages and, to a lesser degree, other 'professional' phagocytes (such as monocytes and dendritic cells) and 'non-professional' phagocytes, such as epithelial cells, conduct efferocytosis. Recent discoveries have shed light on this mechanism and how it works to preserve homeostasis of tissue, repair of tissue and health of the organism. Caspases are a large family of proteases of cysteine acting in cascades. A cascade leading to activation of caspase 3 mediates apoptosis and is responsible for killing cells, hiring macrophages, and presenting a "eat me" signal(s). If macrophages do not effectively engulf apoptotic cells, they undergo secondary necrosis and release intracellular materials that reflect a molecular pattern associated with injury, which can lead to autoimmune diseases. Here, the processes of efferocytosis are illustrated and the pathophysiological effects that which occur when this phase is abrogated are highlighted.

Keywords: apoptosis, efferocytosis, macrophages, caspases, pyroptosis, necroptosis

1. Introduction

In multicellular species, cell death and the efficient clearance of dying cells are fundamental processes that preserve homeostasis. Cells engage in their demise in virtually all physiological and most pathological scenarios through a controlled cascade of signaling events ('regulated' cell death) [1] Damaged or redundant cells die in a regulated manner and are replaced by new cells derived from progenitors of stem cells [2]. Cell death is essential for growth; during mammalian embryogenesis and development, billions of cells are destroyed in order to form new structures and sustain organ function [3, 4]. When addressing pathological events, including tissue damage and infections, significant numbers of cells often die. Cell death must be carefully controlled; extensive damage caused by heat, mechanical compression or osmotic pressure, for example, can cause necrosis of cells, release their intracellular content to the surrounding environment, and contribute to the activation of inflammatory immune pathways that can harm healthy cells and tissues surrounding them. In both homeostasis and disease, removal of cellular corpses is necessary. The phagocyte engulfment of dead cells, a multistep mechanism known as efferocytosis, enables cellular components to be recycled by multicellular organisms. Autoimmune diseases and other diseases may occur when the disposal of cell corpses is faulty. While the degradation and recycling of the mass of a cell are typical features in the clearance of any dead cell, certain cell clearance characteristics are specific to a particular mode of cell death. Depending on their mode of death, dying cells can expose and secrete signals that attract phagocytes, favor their swallowing, or encourage a return to tissue homeostasis. Via regulation of macrophage function following efferocytosis, different forms of cell death may also confer pro-inflammatory or anti-inflammatory signals. There are several cases in which uncoupled phenomena are cell death and corpse disposal, meaning cell death does not follow disposal immediately. The shedding of dead intestinal epithelial cells from the tip of the villus to the intestinal lumen 5 is one such example. Conversely, cells engulf and destroy healthy neighbouring cells in the course of entosis. Although the responsible mechanisms are usually different from those for the removal of dead cells, [5–11] shared aspects have been identified. Nonetheless, dying cells are easily cleared by efferocytosis in the vast majority of cases, and such uncoupled events are not further considered herein. Here we summarise our current understanding of the mechanisms of efferocytosis and how the physiology of the organism is affected by efferocytosis, including inflammation and adaptive immune response results. Every day, several hundred billion cells die and are replaced by newly formed cells. Senescent neutrophils die and are then phagocytized in the bone marrow, liver, and spleen by macrophages [7, 8], while senescent red blood cells are recognized and phagocytized by spleen and liver macrophages [7, 8]. In the small intestine, senescent enterocytes are shed into the intestinal lumen [11]. Cells infected with bacteria or viruses undergo autonomous necrosis of the cells or are destroyed by the immune system. Macrophages as a whole do not recognize these cells, and necrotic cells release materials that may activate the immune system. Inflammation caused by a bacterial or viral infection requires significant white blood cell development, and these cells rapidly die when the infection ceases. Additionally, by inhibiting the synthesis of purines, DNA, or RNA or by cross-linking or intercalating DNA, various anticancer drugs destroy tumour cells.

2. Programmed cell death

Originally, programmed cell death in the sense of insect development was described by Lockshin and Williams [2]. Subsequently, Kerr et al. [12] noted two morphologically distinct forms of cell death in humans, apoptosis and necrosis, through ultrastructural study. The cells swell in necrosis, with the rupture of plasma membranes and release of cellular components; the cells shrink with integral yet ruffling plasma membranes in apoptosis, and nuclei are condensed and fragmented. To avoid the release of intracellular elements, apoptotic cells are rapidly phagocytized by macrophages. This mechanism prevents inflammatory factors from being released and is thus called the death of clean cells. “Programmed cell death” and “apoptosis” are also used synonymously, because apoptosis occurs in developing embryos or in cells that die under physiological conditions. The cell death occurring during inflammation or infection with a necrotic morphology was also found to be programmed or regulated by gene products and was classified as necroptosis and pyroptosis [13]. In addition, non-apoptotic cell death was observed during the development of *Caenorhabditis elegans* [14] and *Drosophila metamorphosis* [15], suggesting that cell death can occur through a non-apoptotic mechanism in animal development. Programmed cell death should therefore not be used as a synonym for apoptosis; it should, as originally described, be reserved for the cell death that takes place in animal development [2].

2.1 Apoptosis

In animal growth, apoptosis plays a significant role. For instance, apoptosis removes interdigital cells, non-functional nerve cells, and activated lymphocytes. Apoptosis also mediates the involution of mammary glands. Apoptosis, does not play a significant role in clearing senescent cells (red blood cells, and intestinal enterocytes). Specific sets of caspases that function in cascades mediate apoptosis, at the end of which caspase 3 or 7 is responsible for killing the cells. In a mechanism called efferocytosis, apoptotic cells are engulfed by macrophages [16]. It is difficult to find free apoptotic cells *in vivo*, even in tissues where large numbers of cells undergo apoptosis, because efferocytosis is effective and rapid [17]. Apoptosis therefore requires pathways not just for destroying cells, but also for recruiting Macrophages ("find me") [18, 19] and providing the macrophages with a signal ("eat me") for cell engulfment [20, 21].

3. Recognition of dying cells

In early studies, cell shrinkage prior to cell death was associated with a process of non-inflammatory cell clearance, while inflammatory response is associated with cell swelling prior to death [12–14] (**Figure 1**). These morphologies were associated with apoptosis and necrosis processes. In recent decades, more molecularly diverse controlled cell death programs have been identified [2], with each type having a major impact on the biological consequences of cell death. Both apoptotic and non-apoptotic dying cells demonstrate and release molecular signals in order to signal phagocytes and direct subsequent phagocytic and immune responses.

4. Efferocytosis

The term "efferocytosis" is derived from the Latin word "efferre" meaning "to take to the grave," which describes the process of programmed cell removal (PrCR). In cell biology, efferocytosis is characterized as the swallowing and decomposition of apoptotic cells by phagocytes. The noteworthy point of efferocytosis is that this mechanism includes the production of anti-inflammatory cytokines and the inhibition of pro-inflammatory cytokines. This means that efferocytosis takes place in an immunologically silent state that mechanistically differs from the phagocytosis of pathogens and other opsonized particles [20] in which pro-inflammatory cytokines are produced. Recent evidence has shown that many types of cells have the ability to clear their neighboring dying cells, including both professional and non-professional phagocytes, although it has long been believed that apoptotic cells can only be cleared by efferocytosis by specialized cells. Professional phagocytes include macrophages and immature dendritic cells, while non-professional phagocytes are involved in epithelial cells, endothelial cells, and fibroblasts. The stages in which apoptotic particles are extracted by professional and non-professional phagocytes are the

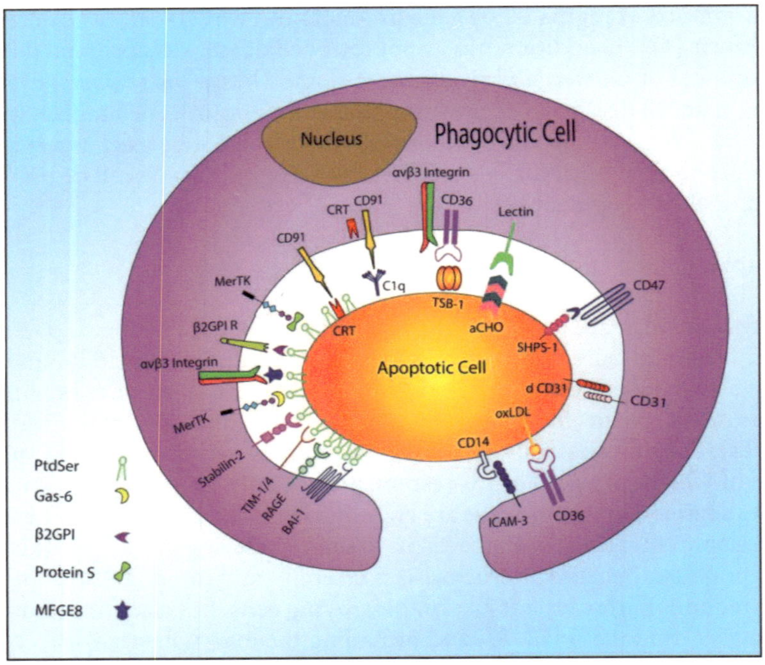

Figure 1.
Recognition of apoptotic cells by phagocytes (The Engulfment Synapse). altered carbohydrate (aCHO), β2-glycoprotein I receptor (β2GPIR), brain-specific angiogenesis inhibitor (BAI-1), complement protein C1q (C1q), calreticulin (CRT), growth arrest-specific 6 (Gas6), intercellular adhesion molecule 3 (ICAM-3), Mer tyrosine kynase (MerTK), milk fat globule-EGF factor 8 protein (MFGE8), oxidized low-density lipoprotein (oxLDL), phosphatidyl serin (PtdSer), receptor for advanced glycation end products (RAGE), Src homology 2 domain-bearing protein tyrosine phosphatase substrate-1 (SHPS-1), T-cell immunoglobulin mucin receptor (TIM), thrombospondin-1 (TSB-1).

same, although various kinetics have been recorded. In fact, the process of efferocytosis is carried out at greater speed and capacity by trained phagocytes [22, 23].

4.1 Find-me signals

Signals produced by apoptotic cells usually involve two types: soluble molecules and complicated extracellular vesicles. Nucleotides such as adenosine triphosphate (ATP) and uridine triphosphate are among the most widely recognized find-me signals (UTP). In the environment, such nucleotides are released by apoptotic cells through the pannexin-1 channel. They bind purinoreceptor 2 (P2Y2) to the surface of the phagocytes, thus serving as short-range chemo attractants [22–27]. Recent studies have shown that certain mediators of oxidative stress can serve as find-me signals by activating monocyte recruitment towards suffering cells at risk of death. The release of oxidative stress mediators may be a physiological mechanism for the precautionary use of phagocytes before the onset of cell death, but it may also promote the pathogenesis of some conditions [1, 9, 28–31]. Lysophosphatidylcholine (LPC) is a lipid mediator formed and released from apoptotic cells by the calcium-independent isoform of phospholipase A22 (iPLA2). LPC plays a part in the recruitment of macrophages by reacting with the G2 accumulation (G2A) receptor. G2A is a G-protein-coupled (GPCR) receptor that plays a role in the regulation of cell cycles, oncogenesis, immunity, and proliferation. G2A is found in macrophages, dendritic cells, neutrophils, mast cells, T lymphocytes, and B lymphocytes. Although awareness of this receptor has so far been restricted, its most important role as an LPC receptor is probably to enable the development of monocyte, neutrophil and lymphocyte recruitment chemo attractants.

Fractalkine (CX3CL1) is a find-me signal which is secreted from apoptotic human B cells in extracellular vesicles. Fractalkine acts as a chemokine which facilitates its recruitment into apoptotic cells by binding to its receptor (CX3CR1) on the surface of the phagocyte 13. Fractalkine also increases the expression of MFGE8, which plays a role as a bridging molecule, in phase 13 of efferocytosis. In addition, apoptotic cells can cause the release of a number of cytokines and chemokines, including MCP-1 and IL-8, involved in the recruitment of monocytes and Tam-Horsfall protein 1 (THP-1) neutrophils, by binding to Fas/CD95 [32].

4.2 Eat-me signals

By binding to their receptors on the surface of phagocytes, eat-me molecules produce intracellular signal cascades, rearranging the cytoskeleton and contributing to swelling. Eat-me signals are usually divided into two groups [33–35]. The first group consists of new emerging molecules, such as phosphatidylserine (PtdSer) or annexin I, on the surface of apoptotic cells. The second category consists of existing molecules changed on the surface of the apoptotic cells [e.g., conversion of the intercellular adhesion molecule (ICAM)-3 to CD31] or various modifications in the plasma membrane components (e.g., loss of plasma membrane asymmetry and increased external negative charges) (**Figure 2**). Among the new evolving molecules on the surface of apoptotic cells, PtdSer is the most popular and well-known eat-me signal. PtdSer is actually located naturally on the inner surface of the plasma

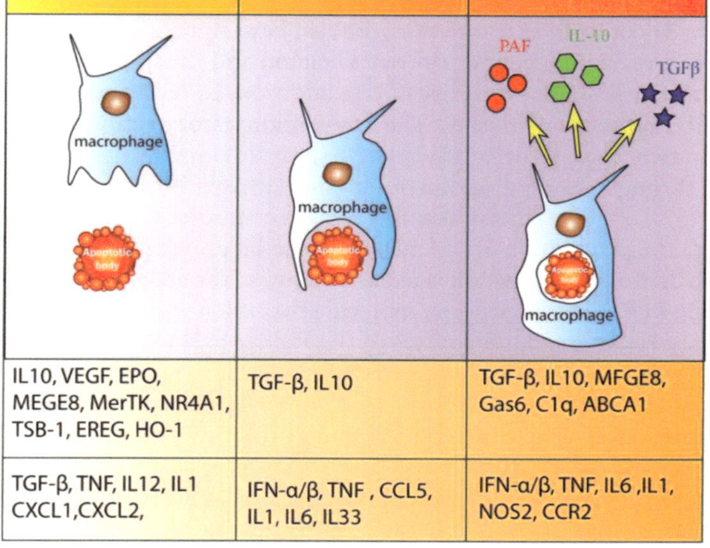

	Find me signaling	Eat me signaling	Post engulfment
Increase	IL10, VEGF, EPO, MEGE8, MerTK, NR4A1, TSB-1, EREG, HO-1	TGF-β, IL10	TGF-β, IL10, MFGE8, Gas6, C1q, ABCA1
Decrease	TGF-β, TNF, IL12, IL1 CXCL1,CXCL2,	IFN-α/β, TNF , CCL5, IL1, IL6, IL33	IFN-α/β, TNF, IL6 ,IL1, NOS2, CCR2

Figure 2.
Apoptotic cell clearance-a multi-stage process. ATP-binding cassette transporter (ABCA1), complement component 1q (C1q), chemokine C-C motif ligand 5 (CCL5), C-C chemokine receptor type 2 (CCR2), chemokine C-X-C motif ligand (CXCL), erythropoietin (EPO), epiregulin (EREG), growth arrest-specific 6 (Gas6), heme oxygenase-1 (HO-1), interferon (IFN), interleukin (IL), Mer tyrosine kinase (MerTK), milk fat globule-EGF factor 8 protein (MFGE8), nitric oxide synthase 2 (NOS2), nuclear receptor subfamily 4 group A (Nr4a1), thrombospondin-1 (TSB-1), transforming growth factor β (TGF-β), tumor necrosis factor (TNF), vascular endothelial growth factor (VEGF).

membrane, but is transmitted to apoptotic cells on the outer surface of the plasma membrane. Flippases are responsible for preserving an asymmetric distribution of phospholipids within the cell membrane in non-apoptotic cells. However, when the apoptosis message is given and caspases in apoptotic cells are activated, flippases become cleaved and inactive. Caspases also simultaneously activate scramblases in apoptotic cells and, as a result, PtdSer rapidly migrates to the outer surface of the plasma membrane [36–40].

PtdSer can bind directly to various surface receptors of phagocytes, including T-cell immunoglobulin mucin (TIM)-1, TIM-4, BAI-1, stabilin-2, and advanced glycation end-product receptors (RAGE). Among them, by binding to PtdSer, TIM-4 is crucial in the tethering of apoptotic cells. Of note, among all phagocytes, some of these receptors do not have the Same distribution [41–46]. Therefore, in the clearance of specific apoptotic cells, each phagocyte with a particular set of these receptors plays a certain function [32, 37–40, 47].

4.3 Bridging molecules

Two domains of receptor-binding have bridging molecules. The prey-binding domain of phagocytosis (PPBD) can, on the one hand, be connected to the surface of apoptotic cells. On the other side, the receptor-binding domain binds to phagocytes. [43, 48–53] A bridge between apoptotic cells and phagocytic cells is thus formed. PtdSer can be indirectly connected to phagocyte receptors by bridging molecules, in addition to its ability to bind directly to receptors on the phagocyte surface. MFGE8 is a bridging molecule which, on the one hand, binds to PtdSer on the surface of apoptotic cells and, on the other, binds to alpha-3/alpha-5 integrins, thus acting as a bridge between apoptotic cells and macrophages in order to facilitate the tethering phase [44, 54–56]. After binding to membrane integrins, it generates signals within macrophages, by recruiting the CRKII-DOCK180-ELMO (**Figure 3**) complex and activating Rac1. These signaling pathways eventually result in a rearrangement of the cytoskeleton and thus, enabling macrophages in the removal of apoptotic cells***. In addition, MFGE8 induces the release of anti-inflammatory cytokines. The anti-inflammatory action of MFGE8 could take place in two ways: directly, through phagocytic capacity development, and indirectly, through the modulation of lipopolysaccharide-activated intracellular signals (LPS) [57–61]. This subset of macrophages promotes tissue repair and angiogenesis through the production of vascular endothelial growth factor (VEGF) or epidermal growth factor (EGF), which is usually found at the stage of tumor promotion. Annexin I (AnxI) is a bridging molecule that is a normal intracellular protein but is transferred to the outer surface of the PtdSer [62] plaque-containing cell membrane during apoptosis.

4.4 Other eat-me signals

While phosphatidylserine is a powerful and well-characterized eat-me signal, other signals may play a role in the identification and swallowing of dying cells [63]. Exposed LPC can bind to IgM on the plasma membrane of dying cells, which in turn binds to phagocyte Fc receptors, such as macrophages [15]. Therefore, LPC continues to serve as a find-me signal as well as an eat-me signal. Proteins present in the endoplasmic reticulum lumen, such as calreticulin, may be exposed to the plasma membrane of dying cells and, in the absence of no-eat-me signals, may serve as an intake-me signal (**Figure 2**). In coordination with complement C1q and mannose-binding lectin (MBL) [16, 17] phagocyte calreticulin is recognized by LDL receptor-related protein 1 (LRP1), also referred to as CD49 [64].

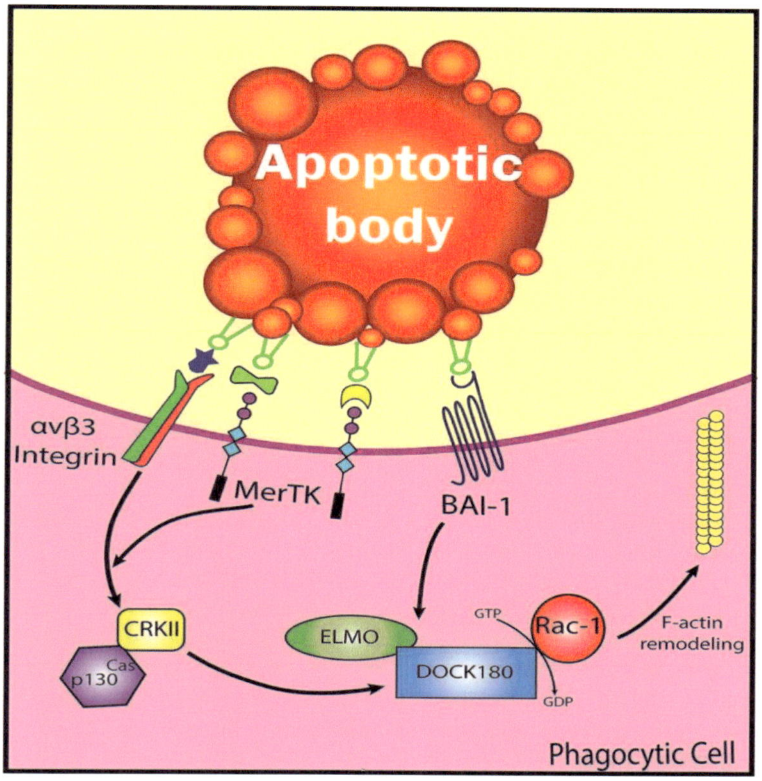

Figure 3.
Activation of the CRKII-DOCK180-ELMO complex within phagocytes. MFGE8-opsonised apoptotic cells are captured by avβ integrins in macrophages; then, the recruitment of the ELMO– DOCK180 complex is induced and finally, by activation of Rac1, cytoskeletal rearrangement occurs for phagocytosis. Brain-specific angiogenesis inhibitor (BAI-1), cellular apoptosis susceptibility protein (Cas), guanosine diphosphate (GDP), guanosine triphosphate (GTP).

4.5 Don't-eat-me signals

Macrophages and other phagocytes are also able to detect live and viable cells through don't-eat-me signals, in addition to their ability to distinguish apoptotic cells from pathogens through eat-me signals. The most important ligand that plays this role is CD47 [65–68]. The signal regulatory protein alpha (SIRP alpha) is the receptor on the surface of macrophages for this protein, the activation of which prevents swallowing even when there is PtdSer [69, 70]. Instead, on the surface of apoptotic cells, CD47 expression is suppressed so that their clearance can be correctly achieved. In transducing don't-eat-me signals, CD31 is another important ligand. Although this protein is not as well-known as CD47, its inhibitory effect on phagocytes [69, 71–73] has been demonstrated by several studies. Another protein that has a critical role in the transduction of a don't-eat-me signal is CD300a. The CD300a-mediated recognition of PtdSer and phosphatidylethanolamine inhibits efferocytosis, preventing swallowing at stage 22.

5. Mechanisms of engulfment

Efferocytosis is a closely regulated process that involves the synchronized swallowing of dead and dying cells, maturation of the phagosome, and then the breakdown of phagolysosomal material. Each stage is regulated by molecular mechanisms that allow the engulfed cell to quickly break down and recover the engulfing phagocyte.

5.1 Uptake of dying cells

If a dying cell is recognized by the phagocyte, swallowing of the cell corpse requires rapid re-organization and synthesis of the plasma membrane to enable the dead cell to be effectively phagocytic internalized. Phagocyte motility and environmental sampling are allowed by a complex mesh of cortical actin fibres lying under the plasma membrane. The phagocyte initiates actin remodeling upon identification of a dying cell, which enables invagination and localised extravagation of the plasma membrane and, eventually, phagosome creation. Depending on the receptor involved, the signaling mechanisms that bridge receptor ligation to activation of actin remodeling and related pathways vary, but usually include coordinated kinase activation (such as those from the SRK, SYK, and protein kinase C families) and phosphatase inactivation, like SHP1, processes that have been well reviewed [20]. Two primary pathways leading to the reorganization of actin occur during efferocytosis and both converge on a central regulator, the small GTPase RAC1 family of RHO. The activation of RAC1 is mediated by LRP1 and the adapter protein GULP96 in the first method, but the exact mechanism of how LRP1 and GULP activate RAC1 is not understood. The second RAC1 activation mechanism is based on the DOCK180 guanine nucleotide exchange factor (GEF) and the phagocytic regulatory protein engulfment and protein motility of cells (ELMO). Another GEF, called 'TRIO', loads GTP onto the small GTPase RHOG following receptor ligation by the dying cell, leading to ELMO [20, 74–76] recruitment. ELMO is then able to interact with DOCK180's SH3 domain [77]. The DOCK180-ELMO complex, together, serves as a GEF for RAC1, leading to activation of RAC1. By activating nucleation-promoting factors of the WASP family, SCAR and WAVE, which recruit the ARP2/3 complex and work together to build an actin nucleation centre, RAC1 is subsequently able to guide localized actin polymerization required to coat or grasp the cargo once triggered through either system [70, 78–80]. The ARP2/3 complex binds existing actin filaments in addition to forming a nucleation center for de novo actin polymerization, enabling new actin synthesis while retaining actin networking and branching, processes that are crucial for the formation of phagosomes [81–86]. Although actin polymerization is a critical component of the creation of phagosomes and the successful capture of the dead cell, actin depolymerization is of equal importance for the plasma membrane splitting of the phagosome. Phagosome sealing processes are highly dependent on different phosphoinositides, including phosphatidylinositol 3,4,5-trisphosphate, which activates GTPase-activating proteins of the RHO family, resulting in the deactivation of GTPases, including RAC1, resulting in depolymerization [87–89]. To facilitate the phagosome separation, a concerted effort exists between actin depolymerization and dynamin. Dynamin 2 also directs the trafficking of the full early phagosome [23, 26] downstream.

5.2 Lysosomal degradation

The phagosome and cell corpse are on their way to a well-planned destructive end after being detected and captured by the dying cell [90]. The phagosome is linked to lysosomes, which include proteases, nucleases, and lipases that degrade the phagosome's cargo. Changes to the phagosome, as discussed below, either trigger or inhibit this fusion. In a multi-stage maturation process, the serine/threonine kinase VPS15 increases the catalytic activity of the early phagosome membrane and the targeting of VPS34 [27, 91]. VPS15 binds to the active RAB5 directly and forms a complex with VPS34 [92]. The role of VPS34 in phagocytosis associated with microtubule-associated protein 1A/1B light chain 3 (LC3) is discussed further (LAP). The transition from early to late phagosomes is characterized by the acquisition of distinct biochemical markers, including small RAB7 GTPase markers, as well as the loss of early RAB5 markers. Late phagosomes are more acidic than early phagosomes due to increased proton pumping into the phagosome lumen mediated by the multimeric protein complex vacuolar ATPase, which translocates H+ across endosomal and phagosomal membranes [36, 47]. RAB7 and its effectors are needed for phagosome maturation, as RAB7 inhibition prevents phagosome-lysosome fusion and results in phagosome acidification failure [32, 91]. RAB7-interacting lysosomal protein (RILP) and oxysterol-binding protein-related protein 1 (ORP1; also known as ORPL1) are found in late endosomes, phagosomes, and lysosomes [32]. By communicating directly with the molecular motor dynein-dynactin, these proteins organize microtubule-dependent vesicular trafficking of RAB7-positive compartments. RILP, dynein, intact microtubules, lysosome-associated membrane protein 1 (LAMP1), and LAMP 2 are involved in the fusion of late phagosomes with the lysosomal compartment [37]. LAMP1 and LAMP2 are highly glycosylated membrane proteins that keep the lysosomal membrane intact and are needed for phagosome-lysosome fusion through interaction with RAB77 [37].

The formation of a Ca2+−dependent SNARE complex, composed of VAMP7 and syntax 7, aids the direct fusion of the mature phagosome with the lysosome [38, 39]. Syntaxine 7 and VAMP7 are both recruited to phagosomes, and phagosome–lysosome fusion is prevented when syntaxine 7 is knocked out [74, 76, 93–98]. The newly developed phagolysosome is distinguished by its high acidity (pH 4.5–5.0) and the presence of active cathepsins after fusion. Acidification is needed for the activation of lysosomal acid hydrolases, which promotes the degradation of the internalized cell body. In conjunction with other effectors such as oxidants, cationic peptides, and lipase, these hydrolytic enzymes mediate the lysosome's potent destructive potential [40, 33].

5.3 Phagosome maturation related modifications

5.3.1 LC3 assisted phagocytosis

The phagosome may be changed to either promote or hinder maturation, affecting the degradation of phagosome material after fusion with lysosomes. During the LAP, one such shift occurs. A subset of proteins necessary for canonical autophagy in LAP, in combination with LAP-specific regulators, allows the LC3 family to be combined with the LAP family. LC3 family proteins facilitate autophagosomes in autophagy, as LC33 depletion also inhibits lysosome fusion [34]. FYCO1, a RAB7 effector protein, interacts with LC3 and PI3P to promote autophagosome-lysosome fusion [35]. Similarly, phagosome LC3 promotes phagosome-lysosome fusion to form a phagolysosome. The rubicon homologue PLEKHM1 interacts with both RAB7 and LC3 and can promote phagosome and endosome trafficking [48, 99]. These processes are most likely shared by autophagia and phagocytosis, both of which are involved in lysosomal targeting and fusion. To date, it has not been thought that the ability to conjugate LC3 with the phagosome directly affects the lysosome's functional potential: impaired maturation and degradation seen in LAP-deficient cells and animals has not been linked to direct lysosomal defects [49, 50].

5.4 Liver X receptors

Other pathways with roles in efferocytosis include the liver X receptors (LXRs) and the PPARs, nuclear receptor families that control genes involved in lipid metabolism and transport [100–103]. By engulfing a dying cell, a phagocyte can effectively double its lipid content, requiring an appropriate transcriptional response which is mediated by these nuclear receptors. LXR deficiency results in their inability to clear the apoptotic thymocytes in macrophages, but does not alter the swallowing of inert beads or other substrates [104]. Similarly, mice that are deficient in LXRs in many tissues accumulate dead cells due to failure of in vivo efferocytosis [104]. Pharmacologic LXR agonists increase efferocytosis and clearance by increasing the expression of receptors, including MERTK, both in cell culture and in vivo136. In addition, during efferocytosis, the existence of LXRs induces the expression of ABCA1 cholesterol transporters, which emit excess cholesterol [105], and genes involved in mitochondrial lipid oxidation, enabling excess lipids to catabolize effectively [105, 106]. Efferocytosis is caused similarly to LXRs by the PPAR family. In response to apoptotic cell engulfment, PPARγ expression occurs and remains elevated until efferocytosis [107] has been resolved. In combination with activation of LXRs, PPARγ activation has been demonstrated to enhance efferocytosis by macrophages [104]. In fact, inhibition of PPARγ by a pharmaceutical antagonist resulted in impaired efferocytosis [107].

5.5 Resolution of phagocytosis

After the phagosome has fused with lysosomes and its cargo has been degraded, a resolution mechanism restores homeostasis within the phagocyte, allowing for further phagocytosis. Since dying cells are not the only phagocytic cargo, the effect of phagocytosis can differ depending on what cargo is internalised. After lysosomal degradation in the form of efferocytosis, some of the components of the cell corpse can be recaptured and recycled for use by the phagocytic cell. Carbohydrates, amino acids, lipids and nucleotides are recycled to replenish cellular stores and can potentially be used as building blocks and sources of energy by the phagocytic cell [108]. The internalisation and degradation of cargo can also influence the activation of signalling pathways; for example, DNA that escapes degradation can activate the DNA sensing cGAS-STING pathway and contribute to the production of type I interferons, as discussed earlier. These events may lead to the loss of the immune tolerance of the apoptotic corpse and can potentially result in autoimmunity [109]. As described earlier, LAP facilitates rapid maturation and generation of phagolysosomes and efficient degradation of cells and promotes the production of anti-inflammatory cytokines, including IL-10, a method that promotes immune silence. On the other hand, LAP disruption decreases phagosome maturation, contributing to the activation of inflammatory signalling pathways and to the production of proinflammatory mediators for IL-1β and [110–113]. The exact mechanism that leads to this switch in cytokine production has not been well known in the absence of LAP. Furthermore, failure in LAP-deficient phagocytes to effectively degrade the cell corpse can result in the leakage of phagosome contents, such as DNA, into the cytosol, which in turn induces the production of type I interferon via STING [114] sensing. STING-dependent interferon expression by tumour-related macrophages in LAP-deficient mice in several cancer models promoted T-cell-mediated anticancer immunity [114]. Following efferocytosis, the phagocyte must restore the functions of cytoskeletal components such as actin and microtubules to ensure that its phagocytic capacity is maximised for additional [115–118]. This restoration of function has not been well studied in contrast with the upstream processes of phagocytosis. However, such details can be inferred from studies investigating the effectors such as RAB27A in the events leading to phagocytosis. RAB27A negatively affects phagocytosis by prolonging the actin coating of nascent phagosomes, resulting in impaired transition to the stage of phagosome sealing [119]. Therefore, RAB27A and related effectors may be essential for the restoration of actin and other cytoskeletal networks following the completion of phagocytosis. It is evident, however, that cytoskeleton restoration after phagocytosis is an ATP-dependent process and may involve multiple components [120]. An additional function of recovery is the recycling or continued expression of membrane receptors to accept subsequent cargo. Impaired recycling results in reduced membrane expression of receptors such as TREM2 and TLR4, which recognize a variety of amyloids and pathogen products [121, 122] respectively, and can also reduce the membrane expression of dying cell recognizing receptors. This decrease in the available surface receptors can impair the endocytic and phagocytic capacity of the phagocyte cell, which means that these restorative events are necessary to ensure that the phagocyte will continue to recognize and clear phagocyte cargo.

6. Disease and efferocytosis

The fact that efferocytosis has so much redundancy in the detection of dying cells emphasizes its significance in pathophysiology. In reality, many autoimmune and inflammatory disorders are associated with defects in this stage, with inflammation, aggravated tissue damage, and organ dysfunction resulting in uncleared or improperly cleared cell bodies. Defects in other aspects of the efferocytotic machinery, on the other hand, could result in enhanced protection against certain diseases.

6.1 Systemic disease

The most common autoimmune disorder associated with compromised efferocytosis is systemic lupus erythematosus (SLE), a chronic systemic autoimmune disease that affects multiple organ systems such as the lungs, kidneys, skin and central nervous system (CNS). Although uncleared dead cells are rarely detected when healthy tissue is examined under the microscope, dead cells can often be observed in the blood, skin, and lymph nodes of SLE patients, and the seriousness of the disease is closely correlated with defective in vitro efferocytosis and dead cell accumulation in vivo [123, 124]. Patients with SLE also exhibit high levels of circulating autoantigens, such as extracellular DNA, that bind autoantibodies to form immune complexes that accumulate or are deposited in the glomerular and vessel walls of the kidneys [125, 126]. Defects at almost every level of the efferocytosis pathway are involved in the pathogenesis of SLE. Mice with genetic deletion that abrogates find-me signalling pathway components (e.g., development of S1P), eat-me signalling pathway components (e.g. TIM4, MFGE8, S protein, MERTK and C1q) and dead cell treatment components (LXRs, PPARs, ABCA1 and rubicon) all display progressive SLE-like disease.

6.2 Neurodegenerative diseases

Phagocytic cells called 'microglia' are present in the CNS. The phenotypically close to these resident phagocytes are the macrophages and clear dead cells and cellular debris in the CNS [127]. However, other CNS cells, such as oligodendrocytes, astrocytes, and neuronal progenitors, can also act as mediators of efferocytosis. Like all organ systems, the CNS needs effective efferocytosis for homeostasis, but efferocytosis is also essential for the reorganization of the neuronal circuits and for the initiation of regenerative response after injury [128]. Consequently, there are numerous neurodegenerative diseases associated with efferocytosis abnormalities, including multiple sclerosis, Alzheimer's disease, Parkinson's disease and Huntington's disease [129, 130]. Multiple sclerosis may result from defects that are involved in cell corpse processing at multiple stages of efferocytosis, including pannexin 1find-me signalling defects, MERTK eat-me signalling, and LXR alpha and LXRβ signalling. Decreased expression of the bridging molecule MFGE8 was also observed in a mouse model of Alzheimer's disease and expression of the dominant negative MFGE8 mutant in microglia cells decreased swallowing of apoptotic neurons [131, 132].

6.3 Lung inflammation

In inflammatory diseases, the production of reactive oxygen species (ROS) by neutrophils is increased. ROS can induce RhoA signalling in phagocytes, thereby reducing the clearance of apoptotic cells through inhibition of efferocytosis. A number of lung inflammatory disorders, including chronic obstructive pulmonary disease (COPD), asthma, pulmonary fibrosis, and cystic fibrosis, may lead to impairments or inefficiencies during inflammation in the effective clearance of dying neutrophils [133].

On the other hand, antioxidants such as N-acetylcysteine (NAC) can increase the clearance of apoptotic cells by inhibiting ROS growth, thereby suppressing RhoA activity and increasing the output of TGF-β [134]. The swallowing and clearance of apoptotic leukocytes can also be enhanced by Macrophages [77, 135–139] by anti-inflammatory drugs such as corticosteroids. In particular, protein S binding to apoptotic cells and MerTK (from the TAM family) upregulation on the surface of macrophages can mediate increased clearance of apoptotic debris induced by corticosteroids.

6.4 Atherosclerosis

Lipid oxidation and retention can cause local inflammation in the arterial wall and encourage atherosclerosis progression. Circulating monocytes can aggregate and differentiate into macrophages in atherosclerotic lesions, which can in turn swallow large amounts of lipids and transform into foam cells filled with fat droplets. Macrophage clearance of apoptotic cells occurs primarily in the intima in the early phases of atherosclerosis, and apoptotic cells are cleared quickly and effectively. The condition is different in the case of advanced atherosclerotic lesions (called plaques), where a large number of monocytes/macrophages are recruited into clear foam cell bodies and additional debris. In foam cell bodies, a large number of oxidised lipids are identified which can inhibit the process of efferocytosis by binding to CD14 and can increase the activity of Rho kinase in atherosclerotic [140, 141]. As a result, macrophages do not effectively complete the process of efferocytosis; they thus transform themselves into foam cells and then promote inflammatory responses by secreting pro-inflammatory cytokines, [109, 121, 142–144] and ROS. The clearance of apoptotic cells also decreases and progression towards secondary necrosis increases, thereby sustaining an unfavorable cycle that leads to atherosclerosis progression and associated complications [145–147].

Therefore, macrophages and the efferocytosis system play a very important role in the development of atherosclerosis [135–139]. The accumulation within the plaques of apoptotic cells and the acceleration of atherosclerotic disease are the result of a defect in the separate mediators of efferocytosis [148]. A study showed, for example, that Gas deficiency contributed to the development of atherosclerosis by reducing the recruitment of lesion phagocytes, while the Mer function defect prevented the clearance of apoptotic cells, which promoted the progression of atherosclerosis. Therefore, a detailed knowledge of the efferocytosis pathways regulated by this cell will open up promising windows for the treatment of atherosclerotic disease because of the critical role of macrophages in atherosclerosis pathogenesis [149].

6.5 Cancer

Although efferocytosis plays a vital role in tissue homeostasis, since clearance of apoptotic cells is followed by immunosuppressive responses, its role in cancer can be detrimental to the host. Indeed, tumour tissue is subject to high-speed cell proliferation and apoptosis, and the process of apoptotic cell clearance, which is associated with anti-inflammatory responses, can encourage tumour escape from immune surveillance and encourage tumour progression. Therefore, some studies have shown that different ligands and receptors involved in efferocytosis may be over-expressed in certain tumour cells and may cause tumorigenesis [40]. Soki et al. have shown that prostate cancer patients' tissue and serum exosomes have a higher degree of MFGE8 expression than controls and that efferocytosis caused by MFGE8 leads to macrophage polarisation towards an M2 phenotype that promotes tumour [33]. Furthermore, in human oral squamous cell carcinoma (SCC) surgical specimens, the immunohistochemical expression of MFGE8 has been associated with some clinical tumour characteristics, including size, stage and invasion patterns [34]. Intensified Mer TK signalling in tumour tissues has been associated with cancer development, metastasis, and resistance to care [35]. Some studies have also been carried out showing that inhibiting the activation of efferocytosis against cancer progression can be successful.

In the co-cultivation of prostatic cancer cells and macrophages, the use of antibodies against MFGE8 substantially decreased the conversion of macrophages to the tumor-promoting M2 phenotype [33]. Tumor production has been minimized in mice by monoclonal antibodies to MFGE8. Consequently, inhibiting efferocytosis and preventing the expression of anti-inflammatory interleukins will improve the immune system's efficacy and anti-tumor effects [99]. However, literature data on the impact of efferocytosis on cancer is troublesome, as there is also evidence that cancer progression may be promoted by a defective efferocytosis. For instance, cancer cells have been shown to express high levels of CD47 on their surface in order to escape from phagocytosis. Therefore, a new approach to cancer treatment, based on inhibiting the expression of CD47 by siRNA or monoclonal antibodies in tumour cells, has emerged to prevent the transduction of don't-eat-me signals into macrophages and increase the phagocytosis of cancer cells [48, 49]. A significant point on this controversial topic is linked to anti-cancer medications. These drugs cause the death of cancer cells by [150–152]. Anti-cancer drug-induced cancer cell apoptosis, accompanied by increased expression of anti-inflammatory cytokines (TGFβ, PGE2, PAF and IL-10), is triggered by efferocytosis in the tumour setting. As a result, this process will lead to the silent anti-tumour response of the immune system [50]. In addition, it has been shown that anthracyclines (e.g., doxorubicin) promote the expression of eat-me signals on the surface of tumour cells, leading to immunogenic cell death [5]. Additionally, as described, during the apoptotic process, apoptotic cells begin to release ATP and UTP within 2 to 4 hours. These nucleoside triphosphates have been demonstrated to act as chemo attractants and to inhibit inflammatory responses in the recruitment of macrophages [11]. The use of chemotherapy drugs to induce apoptosis in tumour cells can increase the release of nucleoside triphosphates up to 100 times higher, with a significant effect on inflammatory inhibition and suppression of the immune system, although the amounts of nucleotides released during apoptosis are low, so that their effects on immune response suppression are limited. Therefore, a serious reconsideration of anti-cancer drugs is anticipated in the coming years, by fully understanding the function of these agents in apoptotic tumour cell efferocytosis [51, 52] and the related therapeutic consequences.

7. Conclusion

From the point of view of molecular biology and its defects in the development of different diseases, this chapter explores the mechanism of efferocytosis. Just a few examples of the potential effect of disrupted efferocytosis on human diseases are cancer, autoimmune diseases, lung inflammation and atherosclerosis. However, impaired clearance of apoptotic cells, including neurodegeneration, skeletal alterations, cardiac growth, electrical impulse propagation and post-injury recovery, angiogenesis and wound healing, is believed to be involved in many other diseases with unfavorable prognosis and processes. In general, in an inflammation-silenced environment, efferocytosis involves many cells and molecular steps leading to the removal of apoptotic cells. Find-me and eat-me signals are secreted by apoptotic cells for this function, while phagocytic cells often work to clear apoptotic debris with the aid of bridging molecules. Anti-inflammatory interleukins are secreted after the engulfment of apoptotic cells. In fact, cancer progression can be increased by both defective and enhanced efferocytosis. Overall, it is possible to inhibit or activate the receptors associated with this process with accurate knowledge of the molecular mechanisms involved in the efferocytosis process and to target strongly for a new window in the treatment of many human diseases.

8. Future perspectives

Efferocytosis is regulated by a number of factors, including complex membrane lipids and multiple effector proteins. These factors are responsible for functions including dead cell recognition, phagocytosis activation, and, finally, cell corpse degradation. The multi-step process from cell death to cell clearance, as previously mentioned, is delicate, demonstrating the importance of efferocytosis in development, homeostasis, and pathophysiology. The ability of phagocytes to clear cellular debris is required for the homeostatic function of nearly every major physiological system. Efferocytosis dysfunction induces serious disease, while activation of efferocytosis encourages immune silence by processing anti-inflammatory signals, clearing apoptotic cells, and involving the LAP machinery. Since current research indicates that efferocytosis and LAP involvement are beneficial in many environments, inducing an inflammatory response in the treatment of various malignancies by inducing LAP and efferocytosis may be beneficial. The production of therapeutics targeting components of these pathways can increase the efficacy of current cancer therapies such as checkpoint blockade by raising the immunogenicity of apoptotic tumor cells. Since efferocytosis and associated molecular mechanisms are still being studied, new insights into the activation and regulation of the dying cell response are likely to lead to the discovery of new cancer, autoimmunity, neurodegeneration, and beyond treatment paradigms. Efferocytosis is regulated by a number of factors that mediate functions such as dead cell identification and activation of the multi-step process from cell death to cell clearance, demonstrating the importance of efferocytosis in growth, homeostasis, and pathophysiology. The ability of phagocytes to clear cellular debris is required, at least in part, for the homeostatic function of almost every major physiological system. Efferocytosis dysfunction causes severe disease, while activation of efferocytosis promotes immune silence through anti-inflammatory signal processing in the context of clearing apoptotic cells and the involvement of the LAP machinery. Since current research suggests that efferocytosis and LAP involvement are beneficial in a variety of settings, inducing an inflammatory response by inducing LAP and efferocytosis can be beneficial in the treatment of various cancers. Established cancer therapies, such as checkpoint blockade, may be improved by developing therapeutics that target components of these pathways, such as enhancing the immunogenicity of apoptotic tumor cells.

Author details

Bichandarkoil Jayaram Pratima and Namasivayam Nalini*
Department of Biochemistry and Biotechnology, Faculty of Science, Annamalai University, Tamilnadu, India

*Address all correspondence to: nalininam@yahoo.co.in

References

[1] Galluzzi, L. et al. Molecular mechanisms of cell death: recommendations of the Nomenclature Committee on Cell Death 2018. *Cell Death Differ.* 25, 486-541 (2018).

[2] Bergmann, A. & Steller, H. Apoptosis, stem cells, and tissue regeneration. *Sci. Signal.* 3, re8 (2010).

[3] Suzanne, M. & Steller, H. Shaping organisms with apoptosis. *Cell Death Differ.* 20, 669-675 (2013).

[4] Fuchs, Y. & Steller, H. Programmed cell death in animal development and disease. *Cell* 147, 742-758 (2011).

[5] Blander, J. M. Death in the intestinal epithelium-basic biology and implications for inflammatory bowel disease. *FEBS J.* 283, 2720-2730 (2016).

[6] Cummings, R. J. et al. Different tissue phagocytes sample apoptotic cells to direct distinct homeostasis programs. *Nature* 539, 565-569 (2016).

[7] Eckhart, L., Lippens, S., Tschachler, E. & Declercq, W. Cell death by cornification. *Biochim. Biophys. Acta* 1833, 3471-3480 (2013).

[8] Florey, O., Krajcovic, M., Sun, Q. & Overholtzer, M. Entosis. *Curr. Biol.* 20, R88–R89 (2010).

[9] Overholtzer, M. et al. A nonapoptotic cell death process, entosis, that occurs by cell-in-cell invasion. *Cell* 131, 966-979 (2007).

[10] Segawa, K. et al. Phospholipid flippases enable precursor B cells to flee engulfment by macrophages. *Proc. Natl Acad. Sci. USA* 115, 12212-12217 (2018).

[11] Metayer, L. E., Vilalta, A., Burke, G. A. A. & Brown, G. C. Anti-CD47 antibodies induce phagocytosis of live, malignant B cells by macrophages via the Fc domain, resulting in cell death by phagoptosis. *Oncotarget* 8, 60892-60903 (2017).

[12] Kerr, J. F., Wyllie, A. H. & Currie, A. R. Apoptosis: a basic biological phenomenon with wide-ranging implications in tissue kinetics. *Br. J. Cancer* 26, 239-257 (1972).

[13] Majno, G. & Joris, I. Apoptosis, oncosis, and necrosis. An overview of cell death. *Am. J. Pathol.* 146, 3-15 (1995).

[14] Wyllie, A. H., Kerr, J. F. & Currie, A. R. Cell death: the significance of apoptosis. *Int. Rev. Cytol.* 68, 251-306 (1980).

[15] Kim, S. J., Gershov, D., Ma, X., Brot, N. & Elkon, K. B. I-PLA(2) activation during apoptosis promotes the exposure of membrane lysophosphatidylcholine leading to binding by natural immunoglobulin M antibodies and complement activation. *J. Exp. Med.* 196, 655-665 (2002).

[16] Gardai, S. J. et al. Cell-surface calreticulin initiates clearance of viable or apoptotic cells through trans-activation of LRP on the phagocyte. *Cell* 123, 321-334 (2005).

[17] Ogden, C. A. et al. C1q and mannose binding lectin engagement of cell surface calreticulin and CD91 initiates macropinocytosis and uptake of apoptotic cells. *J. Exp. Med.* 194, 781-795 (2001).

[18] Green, D. R., Oguin, T. H. & Martinez, J. The clearance of dying cells: table for two. *Cell Death Differ.* 23, 915-926 (2016).

[19] MiksaM, Komura H,Wu R, Shah KG, Wang P. 2009. A novel method to determine the engulfment of apoptotic

cells by macrophages using pHrodo succinimidyl ester. *J. Immunol. Methods* 342:71-77.

[20] Ma, Z. et al. Regulation of Rac1 activation by the low-density lipoprotein receptor-related protein. *J. Cell Biol.* 159, 1061-1070 (2002).

[21] Park, D. et al. BAI1 is an engulfment receptor for apoptotic cells upstream of the ELMO/Dock180/Rac module. *Nature* 450, 430-434 (2007).

[22] Sevajol, M. et al. The C-terminal polyproline-containing region of ELMO contributes to an increase in the life-time of the ELMO-DOCK complex. *Biochimie* 94, 823-828 (2012).

[23] Marie-Anais, F., Mazzolini, J., Herit, F. & Niedergang, F. Dynamin-actin cross talk contributes to phagosome formation and closure. *Traffic* 17, 487-499 (2016).

[24] Elliott, M. R. & Ravichandran, K. S. The dynamics of apoptotic cell clearance. *Dev. Cell* 38, 147-160 (2016).

[25] Waterborg, C. E. J. et al. Protective role of the MER tyrosine kinase via efferocytosis in rheumatoid arthritis models. *Front. Immunol.* 9, 742 (2018).

[26] Nienhuis, H. L. et al. AGE and their receptor RAGE in systemic autoimmune diseases: an inflammation propagating factor contributing to accelerated atherosclerosis. *Autoimmunity* 42, 302-304 (2009).

[27] Flannagan, R. S., Jaumouille, V. & Grinstein, S. The cell biology of phagocytosis. *Annu. Rev. Pathol.* 7, 61-98 (2012).

[28] Liu, H. & Jiang, D. Fractalkine/CX3CR1 and atherosclerosis. *Clin. Chim. Acta* 412, 1180-1186 (2011).

[29] Liu, M., Tso, P. & Woods, S. C. Receptor CD36 links a risk-associated allele to obesity and metabolic disorders. *J. Biol. Chem.* 293, 13349-13350 (2018). Pharmacologic inhibition of integral membrane protein CD36 significantly reduces body weight gain and improves glucose tolerance in animals receiving a high-fat diet.

[30] Boucher, P. & Herz, J. Signaling through LRP1: protection from atherosclerosis and beyond. *Biochem. Pharmacol.* 81, 1-5 (2011).

[31] Darland-Ransom M, Wang X, Sun CL, Mapes J, Gengyo-Ando K, et al. 2008. Role of *C. elegans* TAT-1 protein in maintaining plasma membrane phosphatidylserine asymmetry. *Science* 320:528-31

[32] Lukacs, G. L., Rotstein, O. D. & Grinstein, S. Determinants of the phagosomal pH in macrophages. In situ assessment of vacuolar H+-ATPase activity, counterion conductance, and H+ "leak". *J. Biol. Chem.* 266, 24540-24548 (1991).

[33] Collins, R. F., Schreiber, A. D., Grinstein, S. & Trimble, W. S. Syntaxins 13 and 7 function at distinct steps during phagocytosis. *J. Immunol.* 169, 3250-3256 (2002)

[34] McEwan, D. G. et al. PLEKHM1 regulates autophagosome-lysosome fusion through HOPS complex and LC3/GABARAP proteins. *Mol. Cell* 57, 39-54 (2015).

[35] Tabata, K. et al. Rubicon and PLEKHM1 negatively regulate the endocytic/autophagic pathway via a novel Rab7-binding domain. *Mol. Biol. Cell* 21, 4162-4172 (2010).

[36] Murray, J. T., Panaretou, C., Stenmark, H., Miaczynska, M. & Backer, J. M. Role of Rab5 in the recruitment of hVps34/p150 to the early endosome. *Traffic* 3, 416-427 (2002).

[37] Harrison, R. E., Bucci, C., Vieira, O. V., Schroer, T. A. & Grinstein, S.

Phagosomes fuse with late endosomes and/or lysosomes by extension of membrane protrusions along microtubules: role of Rab7 and RILP. *Mol. Cell Biol.* 23, 6494-6506 (2003).

[38] Huynh, K. K. et al. LAMP proteins are required for fusion of lysosomes with phagosomes. *EMBO J.* 26, 313-324 (2007)

[39] Braun, V. et al. TI-VAMP/VAMP7 is required for optimal phagocytosis of opsonised particles in macrophages. *EMBO J.* 23, 4166-4176 (2004).

[40] Fairn, G. D. & Grinstein, S. How nascent phagosomes mature to become phagolysosomes. *Trends Immunol.* 33, 397-405 (2012).

[41] Kiss, R. S., Elliott, M. R., Ma, Z., Marcel, Y. L. & Ravichandran, K. S. Apoptotic cells induce a phosphatidylserine-dependent homeostatic response from phagocytes. *Curr. Biol.* 16, 2252-2258 (2006).

[42] Yoon, Y. S. et al. PPARgamma activation following apoptotic cell instillation promotes resolution of lung inflammation and fibrosis via regulation of efferocytosis and proresolving cytokines. *Mucosal Immunol.* 8, 1031-1046 (2015).

[43] Cunha, L. D. et al. LC3-Associated phagocytosis in myeloid cells promotes tumor immune tolerance. *Cell* 175, 429-441.e16 (2018). Genetic ablation of LAP in myeloid cells of the tumour microenvironment promotes anticancer immunity in several model systems by the engagement of a type I interferon response.

[44] Yokoyama, K. et al. Rab27a negatively regulates phagocytosis by prolongation of the actin-coating stage around phagosomes. *J. Biol. Chem.* 286, 5375-5382 (2011).

[45] Kuiper, J. W. et al. Creatine kinase-mediated ATP supply fuels actin-based events in phagocytosis. *PLoS Biol.* 6, e51 (2008).

[46] Manfioletti G, Brancolini C, Avanzi G, Schneider C. 1993. The protein encoded by a growth arrestspecific gene (gas6) is a newmember of the vitamin K-dependent proteins related to protein S, a negative coregulator in the blood coagulation cascade. *Mol. Cell. Biol.* 13:4976-85.

[47] Kissing, S. et al. Vacuolar ATPase in phagosome-lysosome fusion. *J. Biol. Chem.* 290, 14166-14180 (2015).

[48] Kidani, Y. & Bensinger, S. J. Liver X receptor and peroxisome proliferator-activated receptor as integrators of lipid homeostasis and immunity. *Immunol. Rev.* 249, 72-83 (2012).

[49] Schulman, I. G. Liver X receptors link lipid metabolism and inflammation. *FEBS Lett.* 591, 2978-2991 (2017).

[50] Zhang, X., Heckmann, B. L., Campbell, L. E. & Liu, J. G0S2: a small giant controller of lipolysis and adipose-liver fatty acid flux. *Biochim. Biophys. Acta Mol. Cell Biol. Lipids* 1862, 1146-1154 (2017).

[51] N, A. G. et al. Apoptotic cells promote their own clearance and immune tolerance through activation of the nuclear receptor LXR. *Immunity* 31, 245-258 (2009).

[52] Rebe, C. et al. Induction of transglutaminase 2 by a liver X receptor/retinoic acid receptor alpha pathway increases the clearance of apoptotic cells by human macrophages. *Circ. Res.* 105, 393-401 (2009).

[53] Fond, A. M., Lee, C. S., Schulman, I. G., Kiss, R. S. & Ravichandran, K. S. Apoptotic cells trigger a membrane-initiated pathway to increase ABCA1. *J. Clin. Invest.* 125, 2748-2758 (2015).

[54] Lucin, K. M. et al. Microglial beclin 1 regulates retromer trafficking and

phagocytosis and is impaired in Alzheimer's disease. *Neuron* 79, 873-886 (2013).

[55] Baumann, I. et al. Impaired uptake of apoptotic cells into tingible body macrophages in germinal centers of patients with systemic lupus erythematosus. *Arthritis Rheum.* 46, 191-201 (2002).

[56] Liu, Z. & Davidson, A. Taming lupus-a new understanding of pathogenesis is leading to clinical advances. *Nat. Med.* 18, 871-882 (2012).

[57] Witting, A., Muller, P., Herrmann, A., Kettenmann, H. & Nolte, C. Phagocytic clearance of apoptotic neurons by Microglia/Brain macrophages in vitro: involvement of lectin-,integrin-,andphosphatidylserine-mediated recognition. *J. Neurochem.* 75, 1060-1070 (2000).

[58] Galloway, D. A., Phillips, A. E. M., Owen, D. R. J. & Moore, C. S. Phagocytosis in the brain: homeostasis and disease. *Front. Immunol.* 10, 790 (2019).

[59] Mattson, M. P. Apoptosis in neurodegenerative disorders. *Nat. Rev. Mol. Cell Biol.* 1, 120-129 (2000).

[60] Heckmann, B. L., Tummers, B. & Green, D. R. Crashing the computer: apoptosis vs. necroptosis in neuroinflammation. *Cell Death Differ.* 26, 41-52 (2019).

[61] Fricker, M. et al. MFG-E8 mediates primary phagocytosis of viable neurons during neuroinflammation. *J. Neurosci.* 32, 2657-2666 (2012).

[62] Berden, J. H. Lupus nephritis. *Kidney Int.* 52, 538-558 (1997).

[63] van Bruggen, M. C. et al. Antigen specificity of anti-nuclear antibodies complexed to nucleosomes determines glomerular basement membrane binding in vivo. *Eur. J. Immunol.* 27, 1564-1569 (1997).

[64] Fuller, A. D. & Van Eldik, L. J. MFG-E8 regulates microglial phagocytosis of apoptotic neurons. *J. Neuroimmune Pharmacol.* 3, 246-256 (2008).

[65] Ferguson, T. A. & Green, D. R. Autophagy and phagocytosis converge for better vision. *Autophagy* 10, 165-167 (2014).

[66] Kim, J. Y. et al. Noncanonical autophagy promotes the visual cycle. *Cell* 154, 365-376 (2013).

[67] Nandrot, E. F. & Dufour, E. M. Mertk in daily retinal phagocytosis: a history in the making. *Adv. Exp. Med. Biol.* 664, 133-140 (2010).

[68] Yang, H. et al. Activation of liver X receptor alleviates ocular inflammation in experimental autoimmune uveitis. *Invest. Ophthalmol. Vis. Sci.* 55, 2795-2804 (2014).

[69] Juncadella, I. J. et al. Apoptotic cell clearance by bronchial epithelial cells critically influences airway inflammation. *Nature* 493, 547-551 (2013).

[70] Ten Kate, M. K. & van der Meer, J. Protein S deficiency: a clinical perspective. *Haemophilia* 14, 1222-1228 (2008).

[71] Fox, S., Leitch, A. E., Duffin, R., Haslett, C. & Rossi, A. G. Neutrophil apoptosis: relevance to the innate immune response and inflammatory disease. *J. Innate Immun.* 2, 216-227 (2010).

[72] Hodge, S., Hodge, G., Scicchitano, R., Reynolds, P. N. & Holmes, M. Alveolar macrophages from subjects with chronic obstructive pulmonary disease are deficient in their ability to phagocytose apoptotic airway epithelial

cells. *Immunol. Cell Biol.* 81, 289-296 (2003).

[73] Cohen, P. L. & Shao, W. H. Gas6/TAM receptors in systemic lupus erythematosus. *Dis. Markers* 2019, 7838195 (2019).

[74] Fadok VA, Bratton DL, Rose DM, Pearson A, Ezekewitz RA, Henson PM. 2000. A receptor for phosphatidy lserine-specific clearance of apoptotic cells. *Nature* 405:85-90

[75] Wang X, Wu YC, Fadok VA, Lee MC, Gengyo-Ando K, et al. 2003. Cell corpse engulfment mediated by *C. elegans* phosphatidylserine receptor through CED-5 and CED-12. *Science* 302:1563-66

[76] Li MO, Sarkisian MR, Mehal WZ, Rakic P, Flavell RA. 2003. Phosphatidylserine receptor is required for clearance of apoptotic cells. *Science* 302:1560-63

[77] Bose J, Gruber AD, Helming L, Schiebe S, Wegener I, et al. 2004. The phosphatidylserine receptor has essential functions during embryogenesis but not in apoptotic cell removal. *J. Biol.* 3:15

[78] Ramirez-Ortiz, Z. G. et al. The scavenger receptor SCARF1 mediates the clearance of apoptotic cells and prevents autoimmunity. *Nat. Immunol.* 14, 917-926 (2013).

[79] Walport, M. J., Davies, K. A. & Botto, M. C1q and systemic lupus erythematosus. *Immunobiology* 199, 265-285 (1998).

[80] Zeng, T. et al. The detection of autoantibodies to ATP-binding cassette transporter A1 and its role in the pathogenesis of atherosclerosis in patients with systemic lupus erythematosus. *Clin. Biochem.* 45, 1342-1346 (2012).

[81] Li, Z. & Weinman, S. A. Regulation of hepatic inflammation via macrophage cell death. *Semin. Liver Dis.* 38, 340-350 (2018).

[82] Bellan, M. et al. Gas6/TAM system: a key modulator of the interplay between inflammation and fibrosis. *Int. J. Mol. Sci*. 20, 5070 (2019).

[83] Elliott, M. R. et al. Unexpected requirement for ELMO1 in clearance of apoptotic germ cells in vivo. *Nature* 467, 333-337 (2010).

[84] Cattaneo, M. The platelet P2 receptors in inflammation. *Hamostaseologie* 35, 262-266 (2015).

[85] Henson, P. M. & Bratton, D. L. Allergy: airway epithelial Rac1 suppresses allergic inflammation. *Curr. Biol.* 23, R104–R106 (2013).

[86] Good, M. E. et al. Pannexin 1 channels as an unexpected new target of the anti-hypertensive drug spironolactone. *Circ. Res.* 122, 606-615 (2018).

[87] Duman, J. G., Tu, Y. K. & Tolias, K. F. Emerging roles of BAI adhesion-GPCRs in synapse development and plasticity. *Neural Plast.* 2016, 8301737 (2016).

[88] Wium, M., Paccez, J. D. & Zerbini, L. F. The dual role of TAM receptors in autoimmune diseases and cancer: an overview. *Cells* 7, 166 (2018).

[89] Binder, M. D. et al. Gas6 deficiency increases oligodendrocyte loss and microglial activation in response to cuprizone-induced demyelination. *J. Neurosci.* 28, 5195-5206 (2008).

[90] Lutz, S. E. et al. Contribution of pannexin1 to experimental autoimmune encephalomyelitis. *PLoS One* 8, e66657 (2013).

[91] Rubino, M., Miaczynska, M., Lippe, R. & Zerial, M. Selective membrane

recruitment of EEA1 suggests a role in directional transport of clathrin-coated vesicles to early endosomes. *J. Biol. Chem.* 275, 3745-3748 (2000).

[92] Vieira, O. V. et al. Distinct roles of class I and class III phosphatidylinositol 3-kinases in phagosome formation and maturation. *J. Cell Biol.* 155, 19-25 (2001).

[93] Nagata S, Suzuki J, Segawa K, Fujii T. 2016. Exposure of phosphatidylserine on the cell surface. *Cell Death Differ*. 23:952-61

[94] Chen Y-Z, Mapes J, Lee E-S, Skeen-Gaar RR, Xue D. 2013. Caspase-mediated activation of *Caenorhabditis elegans* CED-8 promotes apoptosis and phosphatidylserine externalization. *Nat. Commun.* 4:1-9.

[95] Rogers C, Fernandes-Alnemri T, Mayes L, Alnemri D, Cingolani G, Alnemri ES. 2016. Cleavage of DFNA5 by caspase-3 during apoptosis mediates progression to secondary necrotic/pyroptotic cell death. *Nat. Commun.* 8:1-14

[96] Wang Y, Gao W, Shi X, Ding J, Liu W, et al. 2017. Chemotherapy drugs induce pyroptosis through caspase-3 cleavage of a gasdermin. *Nature* 547:99-103

[97] Kawane K, Fukuyama H, Yoshida H, Nagase H, Ohsawa Y, et al. 2003. Impaired thymic development in mouse embryos deficient in apoptotic DNA degradation. *Nat. Immunol.* 4:138-44.

[98] Hanayama R, Tanaka M, Miwa K, Shinohara A, Iwamatsu A, Nagata S. 2002. Identification of a factor that links apoptotic cells to phagocytes. *Nature* 417:182-87

[99] Heckmann, B. L. et al. Liver X receptor alpha mediates hepatic triglyceride accumulation through upregulation of G0/G1 switch gene 2 expression. *JCI Insight* 2, e88735 (2017).

[100] DeKruyff RH, Bu X, Ballesteros A, Santiago C, Chim YLE, et al. 2010. T cell/transmembrane, Ig, and mucin-3 allelic variants differentially recognize phosphatidylserine andmediate phagocytosis of apoptotic cells. *J. Immunol.* 184:1918-30.

[101] Hafizi S, Dahlb¨ack B. 2006. Gas6 and protein S: vitamin K-dependent ligands for the Axl receptor tyrosine kinase subfamily. *FEBS J*. 273: 5231-44.

[102] Walker FJ. 1980. Regulation of activated protein C by a new protein: a possible function for bovine protein S. *J. Biol. Chem.* 255:5521-24.

[103] Walker FJ. 1981. Regulation of activated Protein C by Protein S: the role of phospholipid in Factor Va inactivation. *J. Biol. Chem.* 256:11128-31.

[104] Schneider C, King RM, Philipson L. 1988. Genes specifically expressed at growth arrest of mammalian cells. *Cell* 54:787-93.

[105] Nagata K, Ohashi K, Nakano T, Arita H, Zong C, et al. 1996. Identification of the product of growth arrest-specific gene 6 as a common ligand for Axl, Sky, andMer receptor tyrosine kinases. *J. Biol. Chem.* 271:30022-27

[106] Tsou W-I, Nguyen K-QN, Calarese DA, Garforth SJ, Antes AL, et al. 2014. Receptor tyrosine kinases, TYRO3, AXL, and MER, demonstrate distinct patterns and complex regulation of ligand-induced activation. *J. Biol. Chem.* 289:25750-63.

[107] Lew ED, Oh J, Burrola PG, Lax I, Zag´ orska A, et al. 2014. Differential TAM receptor-ligand phospholipid interactions delimit differential TAM bioactivities. *eLife* 3:e03385.

[108] Thornley TB, Fang Z, Balasubramanian S, Larocca RA, GongW, et al. 2014. Fragile TIM-4–expressing tissue resident macrophages are migratory and immunoregulatory. *J. Clin. Investig.* 124:3443-54.

[109] Bonnardel J, Da Silva C,Henri S, Tamoutounour S, Chasson L, et al. 2015. Innate and adaptive immune functions of Peyer's patch monocyte-derived cells. *Cell Rep*. 11:770-84.

[110] Cao L, ChangH, Shi X, Peng C, He Y. 2016. Keratinmediates the recognition of apoptotic and necrotic cells through dendritic cell receptor DEC205/CD205. *PNAS* 113: 13438-43

[111] Stubbs JD, Lekutis C, Singer KL, Bui A, Yuzuki D, et al. 1990. cDNA cloning of a mouse mammary epithelial cell surface protein reveals the existence of epidermal growth factor-like domains linked to factor VIII-like sequences. *PNAS* 87:8417-21.

[112] Hanayama R, Tanaka M, Miyasaka K, AozasaK,Koike M, et al. 2004. Autoimmune disease and impaired uptake of apoptotic cells in MFG-E8-deficient mice. *Science* 304:1147-50.

[113] Miyasaka K, Hanayama R, Tanaka M, Nagata S. 2004. Expression of milk fat globule epidermal growth factor 8 in immature dendritic cells for engulfment of apoptotic cells. *Eur. J. Immunol.* 34:1414-22.

[114] Nakano T, Ishimoto Y, Kishino J, Umeda M, Inoue K, et al. 1997. Cell adhesion to phosphatidylserine mediated by a product of growth arrest-specific gene 6. *J. Biol. Chem.* 272:29411-14.

[115] Sakamoto K, Fukushima Y, Ito K, Matsuda M, Nagata S, et al. 2016. Osteopontin in spontaneous germinal centers inhibits apoptotic cell engulfment and promotes anti-nuclear antibody production in lupus-prone mice. *J. Immunol.* 197:2177-86.

[116] Nikpay M, Goel A, Won H-H, Hall LM,Willenborg C, et al. 2015. A comprehensive 1,000 Genomes– based genome-wide association meta-analysis of coronary artery disease. *Nat. Genet.* 47:1121-30

[117] Miyanishi M, Tada K, Koike M, Uchiyama Y, Kitamura T, Nagata S. 2007. Identification of Tim4 as a phosphatidylserine receptor. *Nature* 450:435-39.

[118] Freeman GJ, Casasnovas JM, Umetsu DT, DeKruyff RH. 2010. *TIM* genes: a family of cell surface phosphatidylserine receptors that regulate innate and adaptive immunity. *Immunol. Rev.* 235:172-89.

[119] Anderson HA, Maylock CA, Williams JA, Paweletz CP, Shu H, Shacter E. 2002. Serum-derived protein S binds to phosphatidylserine and stimulates the phagocytosis of apoptotic cells. *Nat. Immunol.* 4:87-91

[120] Scott RS, McMahon EJ, Pop SM, Reap EA, Caricchio R, et al. 2001. Phagocytosis and clearance of apoptotic cells is mediated by MER. *Nature* 411:207-11.

[121] Nishi C, Toda S, Segawa K, Nagata S. 2014. Tim4- andMerTK-mediated engulfment of apoptotic cells by mouse resident peritoneal macrophages. *Mol. Cell. Biol.* 34:1512-20.

[122] Cohen PL, Caricchio R, Abraham V, Camenisch TD, Jennette JC, et al. 2002. Delayed apoptotic cell clearance and lupus-like autoimmunity in mice lacking the c-mermembrane tyrosine kinase. *J. Exp.Med.* 196:135-40

[123] Lu Q, Lemke G. 2001. Homeostatic regulation of the immune system by

receptor tyrosine kinases of the Tyro 3 family. *Science* 293:306-11.

[124] Camenisch TD, Koller BH, Earp HS, Matsushima GK. 1999. A novel receptor tyrosine kinase, Mer, inhibits TNF-αproduction and lipopolysaccharide-induced endotoxic shock. *J. Immunol.* 162:3498-503.

[125] Sen P, Wallet MA, Yi Z, Huang Y, Henderson M, et al. 2007. Apoptotic cells induce Mer tyrosine kinase–dependent blockade of NF-κB activation in dendritic cells. *Blood* 109:653-60.

[126] Rothlin CV, Ghosh S, Zuniga EI, Oldstone MBA, Lemke G. 2007. TAM receptors are pleiotropic inhibitors of the innate immune response. *Cell* 131:1124-36.

[127] Bosurgi L, Bernink JH, Delgado Cuevas V, Gagliani N, Joannas L, et al. 2013. Paradoxical role of the proto-oncogene Axl and Mer receptor tyrosine kinases in colon cancer. *PNAS* 110:13091-96.

[128] Park D, Hochreiter-Hufford A, Ravichandran K. 2009. The phosphatidylserine receptor TIM-4 does not mediate direct signaling. *Curr. Biol.* 19:346-51.

[129] Somersan S, Bhardwaj N. 2001. Tethering and tickling: a new role for the phosphatidylserine receptor. *J. Cell Biol.* 155:501-4.

[130] Park D, Tosello-Trampont A-C, Elliott MR, Lu M, Haney LB, et al. 2007. BAI1 is an engulfment receptor for apoptotic cells upstream of the ELMO/Dock180/Rac module. *Nature* 450:430-34.

[131] Hanayama R, Nagata S. 2005. Impaired involution of mammary glands in the absence of milk fat globule EGF factor 8. *PNAS* 102:16886-91.

[132] Nakahashi-Oda C, Tahara-Hanaoka S, Honda S, Shibuya K, Shibuya A. 2012. Identification of phosphatidylserine as a ligand for the CD300a immunoreceptor. *Biochem. Biophys. Res. Commun.* 417:646-50.

[133] Miksa M,Wu R, DongW, Komura H, Amin D, et al. 2009. Immature dendritic cell-derived exosomes rescue septic animals via milk fat globule epidermal growth factor-factor VIII *J. Immunol*. 183:5983-90.

[134] Yamaguchi H, Fujimoto T, Nakamura S, Ohmura K, Mimori T, et al. 2010. Aberrant splicing of the milk fat globule-EGF factor 8 (MFG-E8) gene in human systemic lupus erythematosus. *Eur. J. Immunol.* 40:1778-85.

[135] Chang B, Chen Y, Zhao Y, Bruick RK. 2007. JMJD6 is a histone arginine demethylase. *Science* 318: 444-47.

[136] Kwok J, O'Shea M, Hume DA, Lengeling A. 2017. Jmjd6, a JmjC dioxygenase with many interaction partners and pleiotropic functions. *Front. Genet.* 8:32

[137] Wolf A, Schmitz C, B¨ ottger A. 2007. Changing story of the receptor for phosphatidylserine-dependent clearance of apoptotic cells. *EMBO Rep*. 8:465-69

[138] Schlegel R, Williamson P. 2007. P.S. to PS (phosphatidylserine)–pertinent proteins in apoptotic cell clearance. *Sci. STKE* 2007:pe57

[139] Segawa K, Nagata S. 2015. An apoptotic 'eat me' signal: phosphatidylserine exposure. *Trends Cell Biol*. 25:649-50

[140] Chu, Y. et al. Enhanced synaptic connectivity and epilepsy in C1q knockout mice. *Proc. Natl Acad. Sci. USA* 107, 7975-7980 (2010).

[141] Miksa M,Wu R, DongW, Das P, Yang D, Wang P. 2006. Dendritic

cell-derived exosomes containing milk fat globule epidermal growth factor-factor VIII attenuate proinflammatory responses in sepsis. *Shock* 25:586-93.

[142] Chen, X. et al. Apoptotic engulfment pathway and schizophrenia. *PLoS One* 4, e6875 (2009).

[143] Yanagihashi Y, Segawa K,Maeda R, Nabeshima Y, Nagata S. 2017. Mouse macrophages show different requirements for phosphatidylserine receptor Tim4 in efferocytosis. *PNAS* 114:8800-5.

[144] Ichimura T, Bonventre JV, Bailly V, WeiH,Hession CA, et al. 1998. Kidney injury molecule-1 (KIM-1), a putative epithelial cell adhesion molecule containing a novel immunoglobulin domain, is up-regulated in renal cells after injury. *J. Biol. Chem.* 273:4135-42.

[145] Nakaya M, Watari K, Tajima M, Nakaya T, Matsuda S, et al. 2017. Cardiac myofibroblast engulfment of dead cells facilitates recovery after myocardial infarction. *J. Clin. Investig.* 127:383-40.

[146] Yamaguchi H, Takagi J, Miyamae T, Yokota S, Fujimoto T, et al. 2008. Milk fat globule EGF factor 8 in the serum of human patients of systemic lupus erythematosus. *J. Leuk. Biol.* 83:1300-7.

[147] Peng Y, Elkon KB. 2011. Autoimmunity in MFG-E8-deficient mice is associated with altered trafficking and enhanced cross-presentation of apoptotic cell antigens. *J. Clin. Investig.* 121:2221-41.

[148] Atkin-Smith GK, Poon IK. 2017. Disassembly of the dying: mechanisms and functions. *Trends Cell Biol.* 27:151-62

[149] Vandivier RW, Henson PM, Douglas IS. 2015. Burying the dead. *CHEST* 129:1673-82

[150] Ichimura T, Asseldonk EJ, Humphreys BD, Gunaratnam L, Duffield JS, Bonventre JV. 2008. Kidney injury molecule-1 is a phosphatidylserine receptor that confers a phagocytic phenotype on epithelial cells. *J. Clin. Investig.* 118:1657-68.

[151] Yang L, Brooks CR, Xiao S, Sabbisetti V, YeungMY, et al. 2015. KIM-1–mediated phagocytosis reduces acute injury to the kidney. *J. Clin. Investig.* 125:1620-36.

[152] Nakayama M, Akiba H, Takeda K, Kojima Y, Hashiguchi M, et al. 2009. Tim-3 mediates phagocytosis of apoptotic cells and cross-presentation. *Blood* 113:3821-30.

Section 2

Speciality Topics

Chapter 6

Programmed Cell Death (PCD) in Plant: Molecular Mechanism, Regulation, and Cellular Dysfunction in Response to Development and Stress

Raju Mondal, Sreya Antony, Sovan Roy and Sanjib Kumar Chattopadhyay

Abstract

Programmed cell death (PCD) or apoptosis is a genetically programmed cellular process. Though in the plant, a true caspase system is lacking, still PCD can occur throughout the life cycle at any cell type, tissue, and organ part in response to a wide range of stimuli. Here we have discussed the current understanding of plant PCD in terms of different pathways, cellular dysfunction, regulation, and signaling mechanisms. Our present study discussed how and to what extent PCD is involved in pre-zygotic and post-zygotic plant life cycle and emphasized to what extent PCD modulated in response to abiotic and biotic stress. Additionally, the expression profile of different PCD-associated genes that are modulated by developmental stage, biotic-abiotic stress, cellular metabolites are also elucidated. Hence, this study will be helpful for understanding the molecular and structural instincts of PCD in different stages of plant growth and development, response to biotic/abiotic stimuli, and cellular dysfunction.

Keywords: Plant Development, PCD, Signaling, Stress, Caspase-like, metacaspase-like proteases

1. Introduction

Plant cells can cope up with fluctuating environments by modulation of different cellular processes [1]. Programmed cell death (PCD) is the process in which the cell(s) voluntarily commit suicide on several occasions among which the most prominent reasons are morphogenic or developmental changes and stress responses. In plants and animals, PCD plays significant role during development, structure formation, or removing unwanted tissues [2, 3]. It has been also reported that PCD act as a defense mechanism of an organism to prevent the pathogen spread from the infected cell [4].

In general, PCD is grouped into three broad categories; *viz.* apoptosis, autophagy (vacuolar PCD), and necrosis [5]. The detailed cellular mechanism of apoptosis is well studied in the animal system but current knowledge of the execution processes leading to PCD in plants is very scarce [6, 7]. In general, the fundamental mechanism of apoptosis involves chromatin condensation followed by nuclear fragmentation and the formation of the apoptotic body [8]. Besides that, autophagy involves the formation of vacuoles and rupture of vacuoles that results from the release of hydrolyses enzyme and sudden disappearance of cytoplasm [9]. Such autophagic vacuoles contain degenerating organelles and cytosolic content [10]. Lastly, the process of necrosis is an energy-independent process that results in cytoplasmic swelling and rupture of the plasma membrane [5].

Our present study emphasized on current understanding of plant PCD in different dimensions such as understanding the PCD response towards stress, types of PCDs occurs, cellular-hormones signaling mechanism, and proteases involved in different stress conditions and developmental processes. Additionally, our comparative studies showed that the process of true apoptosis is absent in plants but has several similarities, hence this process is termed apoptosis-like PCD (AL-PCD). The *cys*-protease which is responsible for triggering animal PCD is absent in plants but similar kinds of proteases particularly vacuolar processing enzymes (VPEs) and papain-like cysteine proteases (PLCPs), known as metacaspase play the main roles in PCD [6, 11, 12].

2. Central role of developmental PCD (dPCD) in plant

Inductions of morphogenetic changes are an integral part of the development of an organism. PCD is one of the major fundamental cellular processes that plays crucial role in morphogenetic changes in plant systems [13]. In plants, the occurrence of cell death during development is termed developmental PCD (dPCD). PCD can occur throughout the life cycle at any cell type, tissue, and organ part of the plant (**Table 1**).

Plant Part	Mechanism	Reference
Pollen	ROS mediate	[14]
	Osatg7–1 mediate	[15]
	ROS mediate	[16]
	PERSISTENT TAPETAL CELL2 mediate	[17]
	AtLSD1 deathosome mediate	[18]
Female floral buds	*AMC9*, MeGI, BAG5, AifA, and HSP70 mediate	[19]
Female gametophyte	Auxin efflux	[20]
Fruit	Ca^{2+}-dependent nuclease	[21]
Double Fertilization	ROS mediate	[22]
Vegetative cells	*PrpS-PrsS* modulate PCD vegetative cells	[23]
Zygotic Embryogenesis	—	[24]
Somatic Embryogenesis	K^{+}_{ATP} channel activity	[25]
	Phenol-storing cells	[26]
Embryonic Suspensor Cell	—	[27]

Table 1.
List of important plant parts/sites of PCD in a vascular plant suggested developmental PCD (dPCD) involved cellular differentiation of specific cell types.

Among the two other types of cell death, apoptosis is the most understood type of PCD [5]. In plants, the process of true apoptosis is absent but has a similar kind of programmed cell death termed apoptotic like programmed cell death (AL-PCD). AL-PCD is morphologically and biochemically similar to apoptosis, but due to the structural and functional differences of plant cells, there are some changes in the execution process. The general similarities between apoptosis and AL-PCD include (1) cell shrinkage (2) chromatin condensation, (3) mitochondrial permeabilization and depolarization, (4) cytochrome c release, (5) vacuole leakage and fusion with plasmalemma. In the case of mitochondrial permeabilization, the release of cytochrome *c* leads to the formation of a caspase-like protein termed metacaspase. This in turn results in the release of nuclease and several protease enzymes which finally leads to DNA fragmentation and protein degradation [28]. The fundamental phenomenon of PCD in plants occurred across the whole life cycle depicted in **Figure 1**.

Developmental PCD (dPCD) is triggered by vascular cell death [30]. Vascular cell death is mainly associated with developmental stages includes morphogenesis and senescence. The existence of vascular system in the plant cell is the key difference from an animal cell. Vacuolar cell death in plants resembles autophagy in animals. A large portion of the plant cell is occupied by vacuoles and the presence of these lytic vacuoles plays a major role in plant cell death. This type of cell death is known as vacuolar cell death [30]. The mechanism involved in this vacuolar cell death involves

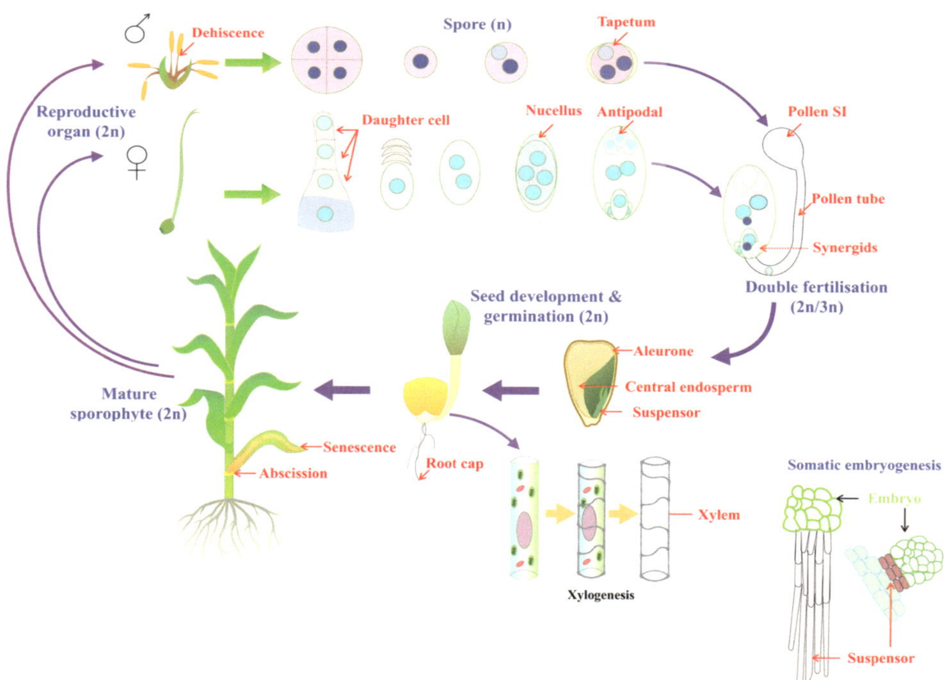

Figure 1.
Central role of PCD across the plant life cycle. Pre-zygotic and post-zygotic developmental changes such as tapetum layer degeneration, daughter cell degeneration, antipodal-synergids cell degeneration, pollen tube degeneration, aleurone layer, central endosperm, suspensor cell death are controlled by PCD. The process of xylogenesis is also influenced by PCD [29]. During the differentiation of tracheary elements, vacuole swelling and rupture are coordinated with the thickening and restructuring of the cell wall. The final collapse of the vacuole immediately precedes nuclear DNA fragmentation, which occurs at the late stages of the cell-death process before the final autolysis of the cell. In mature sporophyte senescence, abscission and dehiscence are regulated by PCD. PCD is also observed during in vitro *somatic embryogenesis [25]. Symplastic cell-to-cell trafficking connection between the somatic embryo and mother explant is broken due to PCD. Degeneration of suspensor cell of somatic embryo mediated by PCD.*

the increase in the size of the vacuoles by fusing small vacuoles and results in the formation of larger vacuoles. Autolytic PCD occurs due to tonoplast rupture and the clearance of cytoplasm. It is not clear the exact reason for the rupture of tonoplast but the clearance of cytoplasm is due to the flow of hydrolases enzyme that degrades the cytoplasm [31, 32] which is released to aid differentiation of tracheary elements [30, 33]. The dPCD involved to develop various organs like integuments [34], megasoprogenesis [35], pollen tube development [36], leaf senescence [37].

Plant cells contain two types of vacuoles in different types of tissues, lytic vacuole (LV) and protein storage vacuole (PSV) [38]. PSV may contain many kinds of proteins especially defense and stress-related, and the pH value found close to neutral [39–41]. Unlike the PSV, LV helps in ion and water homeostasis of the cell [42]. It is reported that PSV of cereal aleurone transformed from storage compartments to lytic organelles and fusion of several PSV, acidification takes place in the vacuolar lumen [43].

Studies also suggest that a particular cysteine protease called the 'vacuolar processing enzyme' (VPE) functions as a key regulator or the executioner of plant vacuolar PCD during development and also during stress [44]. Located in the vacuole, VPE ruptures the lytic vacuolar membrane. VPE is a cysteine protease and involved in cleaves the peptide bond at the C-terminal side of asparagine and aspartic acid [45]. The up-regulation of VPE genes was associated with various types of cell death under stressed conditions. This is essential in processing seed storage proteins and for mediating the susceptible response of toxin-induced cell death [46]. The collapse of the vacuolar membrane allows the release of hydrolytic content and causes the destruction of other organelles [47]. For example degradation of nuclear and chloroplast DNA, without condensation of DNA, can be completed immediately within 20 minutes of vacuole rupture, but chlorophyll degradation takes more time. Even the plasma membrane can be completely degraded as observed in endosperm tissue. However lignified tissues are exempted from this degradation. Furthermore, PCD in tracheary cells and fiber cells is delayed so that more lignification can take place to allow those cells to become more rigid for structural support [47].

Arabidopsis genome codes for at least 4 VPE homologous αVPE, βVPE, γVPE, δVPE. All of which are located in the vacuole [48]. The last one (δVPE) is only found in dicots [49]. Among these βVPE is specifically located to seeds, whereas αVPE and γVPE are specific to vegetative organs [50] and the δVPE is expressed in seed coat development at the early stage of seed formation [51]. However, the number of VPE genes in various plant species differs, for example four genes have been described in Arabidopsis [52], eight in the case of barley [53], 5 in rice [54], and 14 in tomato [55]. In the recent advancement of the genomic and transcriptomic data, the activities of VPE genes along with their expression pattern will become clearer. In an experiment, it has been proved that when all 4 VPE genes are mutated in Arabidopsis (VPE null mutant) and detectable activities of caspase-1 or VPE in the fungal toxin fumonisinB1 (FB1)-treated leaves whereas wild-type leaves had the caspase-1 and VPE activities [46]. It demonstrated that in planta VPE is solely responsible for caspase-1 activity [52].

3. Types of cellular dysfunction and molecular mechanism lead to PCD in plant

In plants, there are mainly three types of PCD that have been reported *viz.* (1) Apoptotic-like cell death (AL-PCD), (2) senescence-associated death, and (3) vacuole-mediated cell death which resembles autophagy [56]. The cellular dysfunction of above mentioned process has been illustrated in **Figure 2**.

Although presence of rigid cell wall associated with cell membrane prevents to form apoptotic body and lack of true caspase as well as phagocytic cells are main reason for absence of true apoptosis in plants but plants exhibit another mechanism that shows striking similarity to apoptosis which is known as apoptotic-like cell death (AL-PCD) [57, 58]. Like animals, when a plant cell is subjected to PCD several changes occur for example cell shrinkage, condensation of chromatin, chromosome fragmentation, mitochondrial permeabilization, cytochrome C release, etc. **Figure 2A** represents these events schematically [59, 60]. Chromatin condensation and chromosome fragmentation are two characteristic features that are also observed during necrotic and autophagic mode cell death.

Hypersensitive reaction (HR) also results of protoplast shrinkage, similar to apoptotic cell shrinkage. But in the case of animal apoptosis, there is a distinct morphology by which apoptosis can be recognized, the plasma membrane retains its integrity, while the cell shrinks. Animal cells form apoptotic bodies -a vesicle containing segments of a dying cell and apoptotic bodies are formed during the execution phase of the apoptotic process, where the cell's cytoskeleton collapse and causes the membrane to bulge outward surrounding cells which might cause damage (like an inflammatory response) to them. However, unlike animal cells, in plant

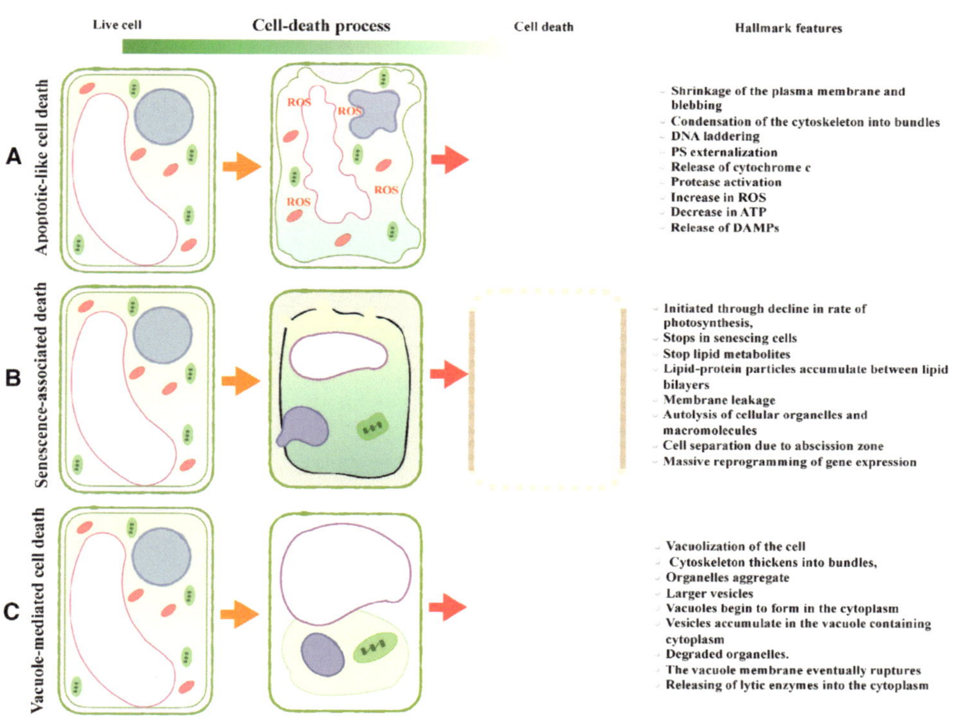

Figure 2.
Morphological comparison, the hallmark features of types of cellular dysfunction in response to PCD. (A) Apoptosis-like PCD mediated by shrinkage of the plasma membrane move away from the cell wall, membrane blebbing, condensation of the cytoskeleton into bundles, DNA laddering, PS externalization, cytochrome c release, protease activation, increase in ROS, decrease in ATP, the release of DAMPs. (B) Senescence-associated death initiated through the decline in the rate of photosynthesis, stops in senescing cells, stops lipid metabolites, lipid-protein particles accumulate between lipid bilayers, causing the membrane to become leaky. In the degenerative phase, autolysis of cellular organelles and macromolecules takes place. And in the terminal phase cell separation takes place at the abscission accumulation by massive reprogramming of gene expression. (C) Vacuole-mediated cell death or autophagic PCD can be characterized by vacuolization of the cell on a large scale. The cytoskeleton thickens into bundles, organelles aggregate, and larger vesicles and vacuoles begin to form in the cytoplasm. Vesicles accumulate in the vacuole containing cytoplasm and degraded organelles. The vacuole membrane eventually ruptures, releasing lytic enzymes into the cytoplasm and furthering cell death.

cells, the content of dead cells remains in the cell itself and there is no membrane blabbing and the process of phagocytosis is also absent [57, 61]. In comparison to animal PCD the true detailed mechanism of AL-PCD is unclear [45]. Plant cells do not undergo 'classic' form of apoptosis because of their rigid cell walls that rule out the necessity or possibility of a breakdown of plant cells into apoptotic bodies and also there are no phagocytic cells in plants [57].

Activation of PCD triggering proteases occurs may be due to a result of a certain change in the cellular environment. It is reported that the activation and dimerization of cysteine C13 protease legumain occur during the low pH. The evidence supporting this includes wheat homolog triticain-α is activated in low pH [62]. In *N. tabacum* reactive carbonyl species (RCS), a ROS product, increased the activity of caspase-like proteases (C1LP and C3LP). This is similar to animal cell where ROS trigger PCD by activating specific proteases [63].

In response to heat stress in cucumber cotyledons, releases of cytochrome c from mitochondria indicate that cyt-c functions differently in plants to initiate PCD in the absence of Bcl-2 proteins [64–66]. The presence of all the cyt-c is

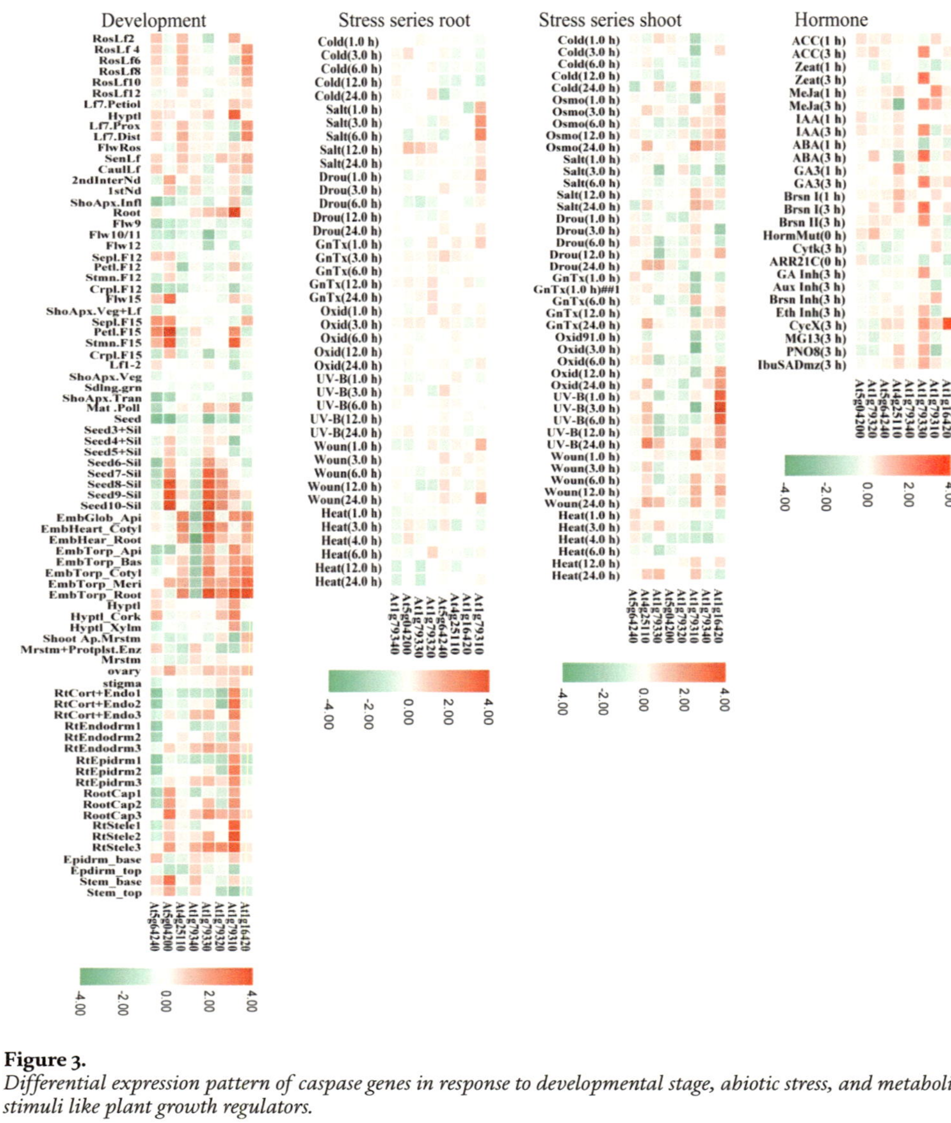

Figure 3.
Differential expression pattern of caspase genes in response to developmental stage, abiotic stress, and metabolic stimuli like plant growth regulators.

detected in the cytoplasm after 1 hour of heat stress although it is completely absent in mitochondria after 3 hours of the heat stress. Self-incompatibility (SI) induction in incompatible pollen tubes is also stimulated by the cytochrome c relocation from mitochondria to cytosol [67]. Like animals, cytochrome c is released in response to stress in plants, but studies suggest that it is functionally different as in the case of apoptosis [60, 64]. Caspase-driven cell death is the process that only present in animal kingdom, but the plant genome lacks core apoptotic proteins like BCL-2 family and caspase [68]. Two caspase-like protein families have been recognized, *viz*. (I) paracaspases, (II) metacaspases [69]. Some metacaspase prodomain comprises a zinc finger motif that resembles the plant hypersensitive response (HR) protein Isd-1 [69]. Metacaspases are members of the C14 class of cysteine proteases and thus related to caspase, orthocaspase, and paracaspase. The metacaspase is recognized as Type I and Type II, both are arginine/lysine-specific, in contrast to caspase, which is aspartate-specific [70]. Differential expression patterns of caspase genes are observed in different developmental stages, abiotic stress, and metabolic stimuli like plant growth regulators (**Figure 3**). Thus, we can assume that plant PCD is a most complex events with coordinated regulation.

Unlike animal cells, the formation of the apoptotic body is absent in plant cells but in response to the biotic and abiotic stress, the apoptotic body-like structures are also observed in the plant cell [71]. According to the most recent proposed model by Thanthrige *et al*. [72], plant PCD is controlled by conserved protein family B cell lymphoma 2 (Bcl-2) associated athanogene (BAG). Subcellular localization and probable function of different BAGs are presented in **Figure 4**.

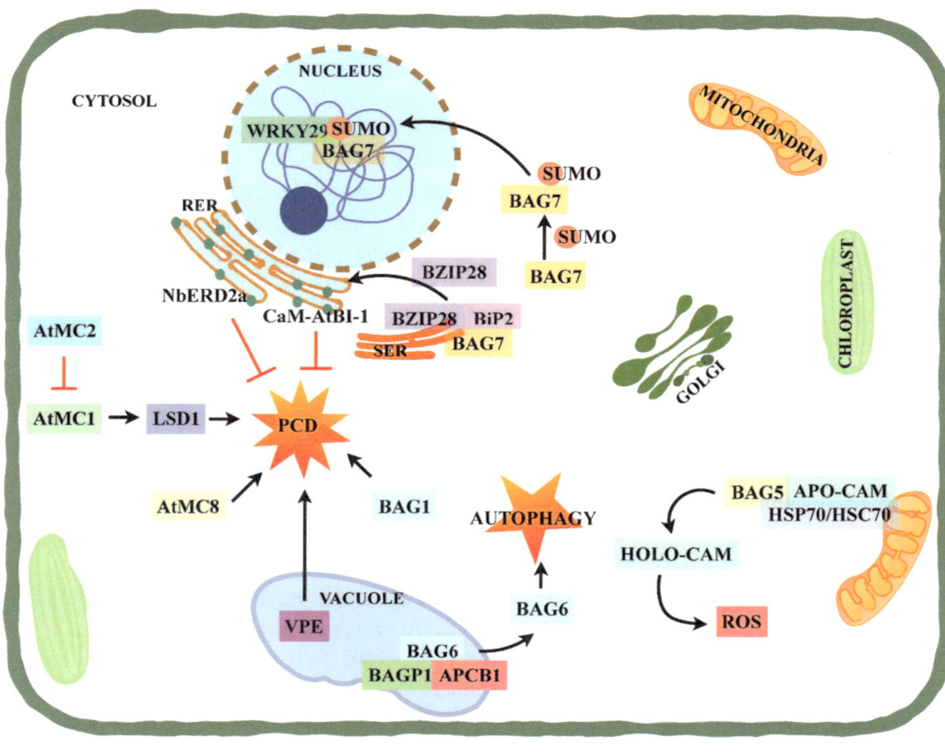

Figure 4.
Pictorial presentation of molecular signature of plant PCD. Seven Arabidopsis BAG members (BAG 1–7) were localized throughout the cell in organelles. AtBAG1–4 (cytoplasm) AtBAG5 (mitochondrion), AtBAG6 (Vacuole, nucleus) and AtBAG7 (ER, nucleus) mechanized PCD in plant system.

In *A. thaliana* there are seven BAG genes identified and they are localized in the cytoplasm, vacuole, nucleus, endoplasmic reticulum (ER), mitochondria [72]. Cytosolic protein AtBAG1, AtBAG2, and AtBAG3 interact at the C-terminus of the HSC70-interacting protein (CHIP). AtBAG1 and AtBAG2 involved proteasomal degradation and plant development process respectively. Though the function of AtBAG3 remains unknown. Cytosolic AtBAG4 interacts with heat-shock protein 70 (HSP70) to repress cell death in response to abiotic stress. AtBAG5 is mitochondrion-localized and interacts with HSP70 and calmodulin (CAM). At a low concentration of cellular Ca^{2+}, the AtBAG5-CAM-HSP70 complex produces reactive oxygen species (ROS) and fasten leaf senescence, but in presence of high cellular Ca^{2+} concentration senescence is inhibited. Vacuole and nucleus localized AtBAG6 play an important role in autophagy. AtBAG6 is bind with AG-associated GRAM protein 1 (BAGP1) and adenomatous polyposis coli B1 (APCB1) are involved in basal defense mechanism against necrotrophic fungi. ER and nucleus localized protein AtBAG7 accelerate heat and cold tolerance by interacting with small ubiquitin-like modifier (SUMO) and WRKY, a DNA-binding protein/transcription factor 29 (WRKY29). Moreover, the BAG co-chaperone family played a potential role in response to a wide range of stress stimulation during plant PCD. Though the future systematic investigation is required to enrich understanding of BAGs function that may help to develop improved stress-tolerant crops.

4. Plant PCD in response to biotic and abiotic stress

Plant PCD occurs mainly during the time of plant-pathogen interaction as well as in response to different abiotic stresses [73]. There are several indentified major causal biotic and abiotic factors that induce plant PCD are presented in **Figure 5**.

Plant-pathogen interaction mediated hypersensitivity response (HR) is a mode of broad-spectrum resistance in plants. Whether the pathogen is compatible or incompatible is determined by the interaction between resistance genes in the host and the avirulence gene in the pathogen as explained in Flor's hypothesis [74]. Hypersensitive cell death is the localized rapid death of cells at the point of infection, and it serves not only to restrict the growth but also stop the invasion of pathogens to other cells/tissue of the plant by undergoing suicide of the infected cells. The complex signalling of this hypersensitive reactions involves cascade of events such as protein phosphorylation, reactive oxygen species (ROS) production, and modification of ion fluxes [75].

HR is known as a defense mechanism against biotrophic plant pathogens or microbes, which depend on nutrients for survival on the host. However, it is observed that HR might be beneficial in the early stages for local adaptation against the infection but not in the later stages [76]. HR is observed in most plant species and it can be influenced by a wide range of plant pathogens such as oomycetes, viruses, fungi, and even insects [77]. All HR-induced PCD have one feature in common which distinguishes it from vacuolar-type, that is, this type of PCD is not initiated by vacuolar swelling and release of lytic enzymes inside plant cell (i.e. non-destructive). However, this type of response may include the release of vacuolar content by fusion with the plasma membrane [78]. Even it has been proved that VPE is essential for a virus-induced HR that involves PCD [11]. Thus it leaves no doubt that vacuolar rupture might take place during HR. Besides that, necrotic cell death is mainly triggered by abiotic stress in plants. The cell death via AL-PCD or necrosis is dependent upon the severity of the stress that the higher heat stress

DOI: http://dx.doi.org/10.5772/intechopen.97940

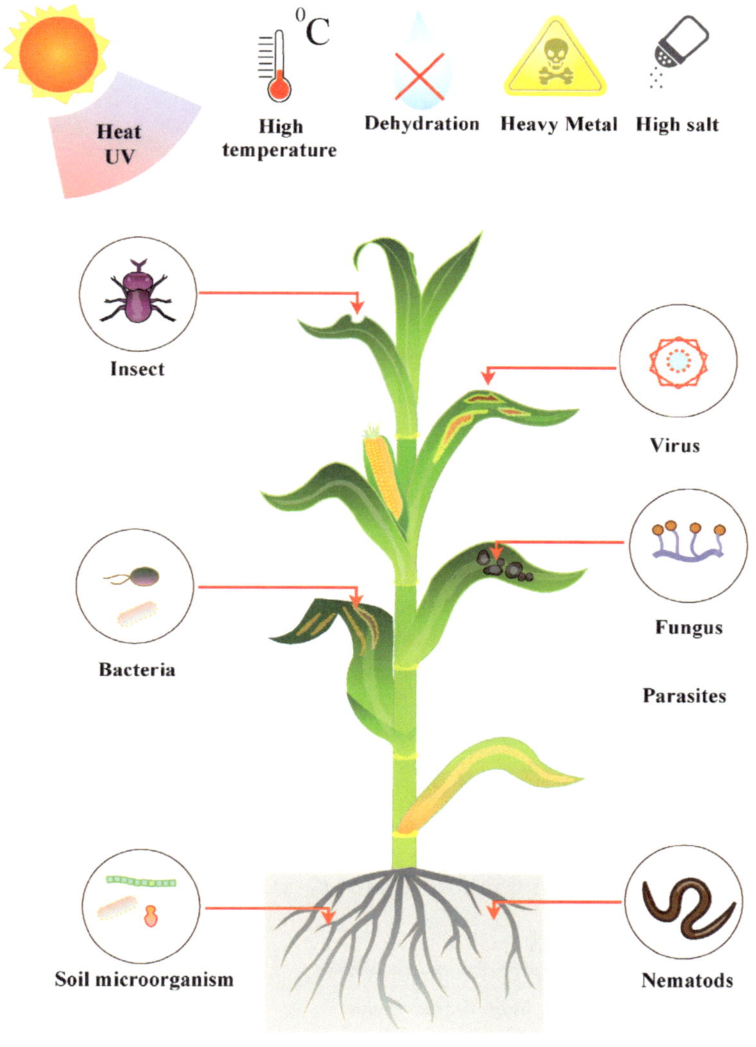

Figure 5.
Diagrammatic presentation of various biotic and abiotic factor-induced PCD in the plant.

leads to necrosis, while moderate heat stress leads to AL-PCD [57]. Previous study also suggests some difference between AL-PCD and necrosis. These include (1) shrinking of protoplast is absent in necrosis while in AL-PCD protoplast shrinks from the cell wall; (2) unlike apoptosis, necrosis is an energy-independent process which proceeds through sudden permeabilization of the plasma membrane and plasma membrane leakage. Additionally, a array of researcher revealed that PCD can induced by a range of abiotic stress tolerance. For more details description see **Table 2**.

5. Metabolic regulation of plant PCD

Hormones play an important role in plant PCD. Ethylene, methyl jasmonate, and salicylate have been identified as key regulators which involved in the

Stress	Studied plants	Cellular responses	Reference
Heat shock	Carrot Arabidopsis Tobacco Soybean	Cytoplasmic condensation	[79–83]
H_2O_2	Arabidopsis	AL-PCD	[79]
Toxin victorin	Oats	Cell shrinkage associated with death	[84]
TMV	Tobacco	DNA Cleavage into 50-kb fragments	[85]
Fungal infection	Cowpea	Internucleosomal cleavage	[86]
D-mannose treatment	Arabidopsis root Maize suspension culture	Oligonucleosomal fragmentation of DNA	[87]
Ultraviolet light or H_2O_2	Metacaspase-8 knockout lines of Arabidopsis	Reduced cell death	[88]
Mycotoxin	Arabidopsis	Detected the presence of Vascular processing enzyme	[52]

Table 2.
List of studies depicted the environmental stress stimulate PCD in plants.

different developmental process like senescence [89]. The increased production of ethylene and salicylic acid associated with ROS-dependent PCD has been investigated [90]. In response to PCD, identified up-regulated genes and associated metabolism processes are autophagy transport, response to ROS, ABA signaling, metal-ion binding, DNA/Protein binding, carotene metabolism, glutamine synthase 2, caspase activity, pectinesterase activity, ethylene signaling, and lipid catabolism. Simultaneously down-regulated genes and associated metabolism processes amino acid metabolism, chlorophyll biosynthesis, carotenoid biosynthesis cytokinin-mediated signaling, glycine metabolism,

PGRs/Metabolites	Identified pathways/mechanism	Plant species	Reference
Auxin	Auxin may restrict the cellulose biosynthesis inhibitor such as Thaxtomin A (TA) and protect PCD.	*Arabidopsis thaliana*	[93]
	Modulate ROS accumulation, anthocyanin production, and the release of mitochondrial cytochrome *c* pathways by auxin antagonist auxinole, auxin inhibitor NPA.	*Aponogeton madagascariensis*	[94]
Nitric oxide	Pathogenesis-related 1 (PR-1) and phenylalanine ammonia-lyase genes and cyclic GMP (cGMP) and cyclic ADP-ribose (cADPR), found to be involved in PCD signaling pathways in abiotic stress.	—	[95]
ROS	*LASTIDIAL NAD-DEPENDENT MALATE DEHYDROGENASE (plNAD-MDH)*, chloroplastic *DICARBOXYLATE TRANSPORTER 1 (DiT1)* and *MITOCHONDRIAL MALATE DEHYDROGENASE 1 (mMDH1)* involved in rescues ROS accumulation and mutant of MOSAIC DEATH 1 (MOD1), leads to the accumulation of ROS and PCD.	*Arabidopsis thaliana*	[96]

PGRs/Metabolites	Identified pathways/mechanism	Plant species	Reference
Salicylic acid (SA)	SA-regulated alternative oxidase (AOX) plays a crucial role in the reduction of mitochondrial ROS and cell death mechanisms.	—	[97]
Ethylene	Ethylene-responsive element binding factors 2 (ERF2) in *Petunia* involved in PCD	*Petunia* sp.	[98]
	Ethylene-mediated ROS signaling plays a role in aerenchyma formation.	*H. annuus*	[99]
Myo-inositol (MI)	MI modulate ROS-induced PCD through SA-dependent and ethylene-dependent pathways.	*Arabidopsis thaliana*	[100]
Phenolic compound	Pyrogallic acid induce PCD in cyanobacteria (*M. aeruginosa) in* caspase-3 (−like)-dependent manner.	*Microcystis aeruginosa*	[101]
Rosmarinic acid	Rosmarinic acid modulates ROS and mitochondrial dysfunction	*Arabidopsis thaliana*	[102]

Table 3.
List of plant growth regulators (PGRs), other metabolites, and their role in plant PCD.

photosynthesis, glutamine synthase 1 [91]. In response to PCD expression profile of ROS-scavenging enzymes such as peroxidase, ascorbate peroxidase (APX), superoxide dismutase (SOD), catalase (CAT), and glutathione peroxidase (GPX)) and non-enzymatic antioxidants ascorbic acid (AsA) and glutathione (GSH) also be studied in *Triticum aestivum* [12]. Yang *et al*, [92] investigated that Acyl-CoA Synthetase (ACOS) is one of the enzymes involved in fatty acids metabolic and protect tapetum cells from PCD and maintain reproductive fitness in rice. Some important findings refers that PGRs, different metabolites are played key role in PCD in plant (**Table 3**). Thus the complex metabolic network is involved in PCD in the plant system. The further systematic metabolic study will be required to understand and enrich our current understanding.

6. Conclusion

In the present study, we have discussed the recent status of occurrence, cellular dysfunction and molecular mechanism of the most complicated fundamental processes of PCD in the plant. Due to the lack of sufficient study on PCD in plant systems, we are unable to identify and describe the exact molecular mechanism. The complex process of PCD is triggered in normal development as well as in response to stress. To understand (1) how plant cell sense or become competent towards PCD? (2) what are the specific signaling pathways correspond to different types of plant PCD?, integrated omic study is helpful in the near future. In our recent study, we have identified novel Cell division cycle and apoptosis regulator 1 (CCAR1) protein in plant system and computationally characterized their function in plant PCD [103]. Thus spectaculars study will be required for understanding the PCD in plant system.

Author details

Raju Mondal[1*], Sreya Antony[1], Sovan Roy[2] and Sanjib Kumar Chattopadhyay[3]

1 Mulberry Tissue Culture Lab, Central Sericultural Germplasm Resources Centre (CSGRC), Hosur, India

2 Santipur College, Santipur, Nadia, West Bengal, India

3 Directorate of Distance Education, Vidyasagar University Midnapore (West), West Bengal, India

*Address all correspondence to: rmcrijaf@yahoo.in

References

[1] Mondal, R. and Das, P., 2020. Data-mining bioinformatics: suggesting *Arabidopsis thaliana* L-type lectin receptor kinase IX. 2 (LecRK-IX. 2) modulate metabolites and abiotic stress responses. *Plant signaling & behavior*, 15(12), 1818031.

[2] Fuchs, Y., & Steller, H. (2011). Programmed cell death in animal development and disease. *Cell*, *147*(4), 742-758.

[3] Van Hautegem, T., Waters, A. J., Goodrich, J., & Nowack, M. K. (2015). Only in dying, life: programmed cell death during plant development. *Trends in plant science*, *20*(2), 102-113.

[4] Lam, E. (2004). Controlled cell death, plant survival and development. *Nature reviews Molecular cell biology*, *5*(4), 305-315.

[5] Kroemer, G., Galluzzi, L., Vandenabeele, P., Abrams, J., Alnemri, E. S., Baehrecke, E. H., ... & Melino, G. (2009). Classification of cell death: recommendations of the Nomenclature Committee on Cell Death 2009. *Cell death & differentiation*, *16*(1), 3-11.

[6] Sychta, K.; Słomka, A.; Kuta, E., 2021. Insights into Plant Programmed Cell Death Induced by Heavy Metals—Discovering aTerra Incognita. Cells, 10, 65.https://doi.org/10.3390/cells10010065

[7] Letai, A. (2017). Apoptosis and Cancer. Annu. Rev. Cancer Biol. 1: 275-294

[8] Adrain, C., & Martin, S. J. (2001). The mitochondrial apoptosome: a killer unleashed by the cytochrome seas. *Trends in biochemical sciences*, *26*(6), 390-397.

[9] Klionsky, D. J., & Emr, S. D. (2000). Autophagy as a regulated pathway of cellular degradation. *Science*, *290*(5497), 1717-1721.

[10] Gozuacik, D., & Kimchi, A. (2007). Autophagy and cell death. *Current topics in developmental biology*, *78*, 217-245.

[11] Hatsugai, N., Kuroyanagi, M., Yamada, K., Meshi, T., Tsuda, S., Kondo, M., Nishimura, M. and Hara-Nishimura, I. (2004). A plant vacuolar protease, VPE, mediates virus-induced hypersensitive cell death. *Science*, *305*(5685), 855-858

[12] Liu, Z., Shi, X., Yan, P., Duan, Y., Geng, X., Ye, J., Li, S., Yang, X., Zhang, G., Jia, Y. and Zhang, L., 2017. Tapetal programmed cell death, antioxidant response and oxidative stress in wheat anthers associated with D2-type cytoplasmic male-sterility. *Scientia Agricultura Sinica*, *50*(21), 4071-4086.

[13] Valandro, F., Menguer, P.K., Cabreira-Cagliari, C., Margis-Pinheiro, M. and Cagliari, A., 2020. Programmed cell death (PCD) control in plants: New insights from the *Arabidopsis thaliana* deathosome. Plant Science, 110603.

[14] Yu, S.X., Feng, Q.N., Xie, H.T., Li, S. and Zhang, Y., 2017. Reactive oxygen species mediate tapetal programmed cell death in tobacco and tomato. *BMC plant biology*, *17*(1), 1-12.

[15] Hanamata, S., Sawada, J., Ono, S., Ogawa, K., Fukunaga, T., Nonomura, K.I., Kimura, S., Kurusu, T. and Kuchitsu, K., 2020. Impact of autophagy on gene expression and tapetal programmed cell death during pollen development in rice. *Frontiers in plant science*, *11*, 172.

[16] Ren, R., Li, Z., Zhou, H., Zhang, L., Jiang, X. and Liu, Y., 2020. Changes in apoptosis-like programmed cell death and viability during the cryopreservation of pollen from *Paeonia*

suffruticosa. *Plant Cell, Tissue and Organ Culture (PCTOC)*, *140*(2), 357-368.

[17] Uzair, M., Xu, D., Schreiber, L., Shi, J., Liang, W., Jung, K.H., Chen, M., Luo, Z., Zhang, Y., Yu, J. and Zhang, D., 2020. PERSISTENT TAPETAL CELL2 is required for normal tapetal programmed cell death and pollen wall patterning. *Plant physiology*, *182*(2), 962-976.

[18] Valandro, F., Menguer, P.K., Cabreira-Cagliari, C., Margis-Pinheiro, M. and Cagliari, A., 2020. Programmed cell death (PCD) control in plants: New insights from the *Arabidopsis thaliana* deathosome. *Plant Science*, 110603.

[19] Wang, Y., Ye, H., Bai, J. and Ren, F., 2020. The regulatory framework of developmentally programmed cell death in floral organs: A review. *Plant Physiology and Biochemistry.*

[20] Wang, J., Guo, X., Xiao, Q., Zhu, J., Cheung, A.Y., Yuan, L., Vierling, E. and Xu, S., 2020. Auxin efflux controls orderly nucellar degeneration and expansion of the female gametophyte in Arabidopsis. *New Phytologist*.

[21] Bai, M., Liang, M., Huai, B., Gao, H., Tong, P., Shen, R., He, H. and Wu, H., 2020. Ca2+-dependent nuclease is involved in DNA degradation during the formation of the secretory cavity by programmed cell death in fruit of *Citrus grandis* 'Tomentosa'. *Journal of experimental botany*, *71*(16), 4812-4827.

[22] Sankaranarayanan, S., Ju, Y. and Kessler, S.A., 2020. Reactive Oxygen Species as Mediators of Gametophyte Development and Double Fertilization in Flowering Plants. *Frontiers in Plant Science*, *11*, 1199.

[23] Lin, Z., Xie, F., Trivino, M., Karimi, M., Bosch, M., Franklin-Tong, V.E. and Nowack, M.K., 2020. Ectopic expression of a self-incompatibility module triggers growth arrest and cell death in vegetative cells. *Plant physiology*, *183*(4), 1765-1779.

[24] Vuosku, J. and Sutela, S., 2020. Microscopical Detection of Cell Death Processes During Scots Pine Zygotic Embryogenesis. In *Plant Embryogenesis* (223-237). Humana, New York, NY.

[25] Petrussa, E., Bertolini, A., Casolo, V., Krajňáková, J., Macrì, F. and Vianello, A., 2009. Mitochondrial bioenergetics linked to the manifestation of programmed cell death during somatic embryogenesis of *Abies alba*. *Planta*, *231*(1), 93-107.

[26] Steinmacher, D.A., Guerra, M.P., Saare-Surminski, K. and Lieberei, R., 2011. A temporary immersion system improves in vitro regeneration of peach palm through secondary somatic embryogenesis. *Annals of Botany*, *108*(8), 1463-1475.

[27] Shi, C., Luo, P., Zhao, P. and Sun, M.X., 2020. Detection of Embryonic Suspensor Cell Death by Whole-Mount TUNEL Assay in Tobacco. *Plants*, *9*(9), 1196.

[28] Connolly, P.F., Jäger, R. and Fearnhead, H.O., 2014. New roles for old enzymes: killer caspases as the engine of cell behavior changes. *Frontiers in physiology*, *5*, p.149.

[29] Iakimova, E.T. and Woltering, E.J., 2017. Xylogenesis in zinnia (*Zinnia elegans*) cell cultures: unravelling the regulatory steps in a complex developmental programmed cell death event. *Planta*, *245*(4), 681-705.

[30] Jones, A. M. (2001). Programmed cell death in development and defense. *Plant Physiology*, *125*(1), 94-97.

[31] Van Doorn, W. G., & Woltering, E. J. (2005). Many ways to exit? Cell death categories in plants. *Trends in plant science*, *10*(3), 117-122.

[32] Avci, U., Earl Petzold, H., Ismail, I. O., Beers, E. P., & Haigler, C. H. (2008). Cysteine proteases XCP1 and XCP2 aid micro-autolysis within the intact central vacuole during xylogenesis in Arabidopsis roots. *The Plant Journal*, *56*(2), 303-315.

[33] Fukuda, H. (2000). Programmed cell death of tracheary elements as a paradigm in plants. In *Programmed Cell Death in Higher Plants* (pp. 1-9). Springer, Dordrecht.

[34] Domínguez, F., & Cejudo, F. J. (2014). Programmed cell death (PCD): an essential process of cereal seed development and germination. *Frontiers in Plant Science*, *5*, 366.

[35] Papini, A., Mosti, S., Milocani, E., Tani, G., Di Falco, P., & Brighigna, L. (2011). Megasporogenesis and programmed cell death in Tillandsia (Bromeliaceae). *Protoplasma*, *248*(4), 651-662.

[36] Wilkins, K. A., Bosch, M., Haque, T., Teng, N., Poulter, N. S., & Franklin-Tong, V. E. (2015). Self-incompatibility-induced programmed cell death in field poppy pollen involves dramatic acidification of the incompatible pollen tube cytosol. *Plant physiology*, *167*(3), 766-779.

[37] Del Duca, S., Serafini-Fracassini, D., & Cai, G. (2014). Senescence and programmed cell death in plants: polyamine action mediated by transglutaminase. *Frontiers in Plant Science*, *5*, 120.

[38] Paris, N., Stanley, C. M., Jones, R. L., & Rogers, J. C. (1996). Plant cells contain two functionally distinct vacuolar compartments. *Cell*, *85*(4), 563-572.

[39] Jiang, L., Phillips, T. E., Hamm, C. A., Drozdowicz, Y. M., Rea, P. A., Maeshima, M., ... & Rogers, J. C. (2001). The protein storage vacuole: a unique compound organelle. *The Journal of Cell Biology*, *155*(6), 991-1002.

[40] Jauh, G. Y., Phillips, T. E., & Rogers, J. C. (1999). Tonoplast intrinsic protein isoforms as markers for vacuolar functions. *The Plant Cell*, *11*(10), 1867-1882.

[41] Okita, T. W., & Rogers, J. C. (1996). Compartmentation of proteins in the endomembrane system of plant cells. *Annual review of plant biology*, *47*(1), 327-350.

[42] Viotti, C. (2014). ER and vacuoles: never been closer. *Frontiers in plant science*, *5*, 20.

[43] Zheng, Y., Zhang, H., Deng, X., Liu, J. and Chen, H., 2017. The relationship between vacuolation and initiation of PCD in rice (*Oryza sativa*) aleurone cells. Scientific reports, 7(1), 1-14.

[44] Hara-Nishimura, I., Hatsugai, N., Nakaune, S., Kuroyanagi, M., & Nishimura, M. (2005). Vacuolar processing enzyme: an executor of plant cell death. *Current opinion in plant biology*, *8*(4), 404-408.

[45] Hatsugai, N., Kuroyanagi, M., Nishimura, M. and Hara-Nishimura, I., 2006. A cellular suicide strategy of plants: vacuole-mediated cell death. *Apoptosis*, *11*(6), 905-911.

[46] Camagna, M. and Takemoto, D., 2018. Hypersensitive response in plants. *eLS*, 1-7.

[47] Obara, K., Kuriyama, H., & Fukuda, H. (2001). Direct evidence of active and rapid nuclear degradation triggered by vacuole rupture during programmed cell death in Zinnia. *Plant physiology*, *125*(2), 615-626.

[48] Zauner, F. B., Dall, E., Regl, C., Grassi, L., Huber, C. G., Cabrele, C., & Brandstetter, H. (2018). Crystal structure of plant legumain reveals a

unique two-chain state with pH-dependent activity regulation. *The Plant Cell*, *30*(3), 686-699.

[49] Yamada, K., Shimada, T., Nishimura, M., & Hara-Nishimura, I. (2005). A VPE family supporting various vacuolar functions in plants. *Physiologia Plantarum*, *123*(4), 369-375.

[50] Kinoshita, T., Yamada, K., Hiraiwa, N., Kondo, M., Nishimura, M., & Hara-Nishimura, I. (1999). Vacuolar processing enzyme is up-regulated in the lytic vacuoles of vegetative tissues during senescence and under various stressed conditions. *The Plant Journal*, *19*(1), 43-53.

[51] Nakaune, S., Yamada, K., Kondo, M., Kato, T., Tabata, S., Nishimura, M., & Hara-Nishimura, I. (2005). A vacuolar processing enzyme, δVPE, is involved in seed coat formation at the early stage of seed development. *The Plant Cell*, *17*(3), 876-887.

[52] Kuroyanagi, M., Yamada, K., Hatsugai, N., Kondo, M., Nishimura, M., & Hara-Nishimura, I. (2005). Vacuolar processing enzyme is essential for mycotoxin-induced cell death in *Arabidopsis thaliana*. *Journal of Biological Chemistry*, *280*(38), 32914-32920.

[53] Julián, I., Gandullo, J., Santos-Silva, L. K., Diaz, I., & Martinez, M. (2013). Phylogenetically distant barley legumains have a role in both seed and vegetative tissues. *Journal of experimental botany*, *64*(10), 2929-2941.

[54] Hatsugai, N., Yamada, K., Goto-Yamada, S., & Hara-Nishimura, I. (2015). Vacuolar processing enzyme in plant programmed cell death. *Frontiers in plant science*, *6*, 234.

[55] Wang, W., Cai, J., Wang, P., Tian, S., & Qin, G. (2017). Post-transcriptional regulation of fruit ripening and disease resistance in tomato by the vacuolar protease SlVPE3. *Genome biology*, *18*(1), 1-23.

[56] Watanabe, N., & Lam, E. (2011). Arabidopsis metacaspase 2d is a positive mediator of cell death induced during biotic and abiotic stresses. *The Plant Journal*, *66*(6), 969-982.

[57] Van Doorn, W.G., Beers, E.P., Dangl, J.L., Franklin-Tong, V.E., Gallois, P., Hara-Nishimura, I., Jones, A.M., Kawai-Yamada, M., Lam, E., Mundy, J. and Mur, L.A.J., 2011. Morphological classification of plant cell deaths. Cell Death & Differentiation, 18(8),1241-1246.

[58] Danon, A., Delorme, V., Mailhac, N., & Gallois, P. (2000). Plant programmed cell death: a common way to die. *Plant Physiology and Biochemistry*, *38*(9), 647-655.

[59] Reape, T. J., & McCabe, P. F. (2010). Apoptotic-like regulation of programmed cell death in plants. *Apoptosis*, *15*(3), 249-256.

[60] Yao, N., Tada, Y., Park, P., Nakayashiki, H., Tosa, Y., & Mayama, S. (2001). Novel evidence for apoptotic cell response and differential signals in chromatin condensation and DNA cleavage in victorin-treated oats. *The Plant Journal*, *28*(1), 13-26.

[61] Reed, J.C., 2000. Mechanisms of apoptosis. The American journal of pathology, 157(5), 1415-1430.

[62] Savvateeva, L. V., Gorokhovets, N. V., Makarov, V. A., Serebryakova, M. V., Solovyev, A. G., Morozov, S. Y., ... & Aliev, G. (2015). Glutenase and collagenase activities of wheat cysteine protease Triticain-α: feasibility for enzymatic therapy assays. *The international journal of biochemistry & cell biology*, *62*, 115-124.

[63] Biswas, M. S., & Mano, J. I. (2016). Reactive carbonyl species activate

caspase-3-like protease to initiate programmed cell death in plants. *Plant and Cell Physiology*, *57*(7), 1432-1442.

[64] Balk, J., Leaver, C. J., & McCabe, P. F. (1999). Translocation of cytochrome c from the mitochondria to the cytosol occurs during heat-induced programmed cell death in cucumber plants. *FEBS letters*, *463*(1-2), 151-154.

[65] Yu, X. H., Perdue, T. D., Heimer, Y. M., & Jones, A. M. (2002). Mitochondrial involvement in tracheary element programmed cell death. *Cell Death & Differentiation*, *9*(2), 189-198.

[66] Martínez-Fábregas, J., Díaz-Moreno, I., González-Arzola, K., Janocha, S., Navarro, J. A., Hervás, M., ... & Miguel, Á. (2013). New *Arabidopsis thaliana* cytochrome c partners: a look into the elusive role of cytochrome c in programmed cell death in plants. *Molecular & Cellular Proteomics*, *12*(12), 3666-3676.

[67] Thomas, S. G., & Franklin-Tong, V. E. (2004). Self-incompatibility triggers programmed cell death in Papaver pollen. *Nature*, *429*(6989), 305-309.

[68] Koonin, E. V., & Aravind, L. (2002). Origin and evolution of eukaryotic apoptosis: the bacterial connection. *Cell Death & Differentiation*, *9*(4), 394-404.

[69] Uren, A.G., O'Rourke, K., Aravind, L.A., Pisabarro, M.T., Seshagiri, S., Koonin, E.V. and Dixit, V.M., 2000. Identification of paracaspases and metacaspases: two ancient families of caspase-like proteins, one of which plays a key role in MALT lymphoma. *Molecular cell*, *6*(4),961-967.

[70] Vercammen, D., Van De Cotte, B., De Jaeger, G., Eeckhout, D., Casteels, P., Vandepoele, K., & Van Breusegem, F. (2004). Type II metacaspases Atmc4 and Atmc9 of *Arabidopsis thaliana* cleave substrates after arginine and lysine. *Journal of Biological Chemistry*, *279*(44), 45329-45336.

[71] Li, M. G., Villemur, R., Hussey, P. J., Silflow, C. D., Gantt, J. S., & Snustad, D. P. (1993). Differential expression of six glutamine synthetase genes in *Zea mays*. *Plant molecular biology*, *23*(2), 401-407.

[72] Thanthrige, N., Jain, S., Bhowmik, S. D., Ferguson, B. J., Kabbage, M., Mundree, S., & Williams, B. (2020). Centrality of BAGs in plant PCD, stress responses, and host defense. *Trends in Plant Science*. S1360-1385(20)30127-8

[73] Burke, R., Schwarze, J., Sherwood, O.L., Jnaid, Y., McCabe, P.F. and Kacprzyk, J., 2020. Stressed to death: The role of transcription factors in plant programmed cell death induced by abiotic and biotic stimuli. *Frontiers in plant science*, *11*, 1235.

[74] Flor, H. H. (1971). Current status of the gene-for-gene concept. *Annual review of phytopathology*, *9*(1), 275-296.

[75] Pontier, D., Balagué, C., & Roby, D. (1998). The hypersensitive response. A programmed cell death associated with plant resistance. *Comptes Rendus de l'Académie des Sciences-Series III-Sciences de la Vie*, *321*(9), 721-734.

[76] Münch, S., Lingner, U., Floss, D. S., Ludwig, N., Sauer, N., & Deising, H. B. (2008). The hemibiotrophic lifestyle of Colletotrichum species. *Journal of plant physiology*, *165*(1), 41-51.

[77] Balint-Kurti, P. (2019). The plant hypersensitive response: concepts, control and consequences. *Molecular plant pathology*, *20*(8), 1163-1178.

[78] Dangl, J. L., & Jones, J. D. (2001). Plant pathogens and integrated defence responses to infection. *nature*, *411*(6839), 826-833.

[79] McCabe, P. F., Levine, A., Meijer, P. J., Tapon, N. A., & Pennell, R. I. (1997).

A programmed cell death pathway activated in carrot cells cultured at low cell density. *The Plant Journal*, *12*(2), 267-280.

[80] McCabe, P. F., & Leaver, C. J. (2000). Programmed cell death in cell cultures. In *Programmed Cell Death in Higher Plants* (pp. 115-124). Springer, Dordrecht.

[81] Vacca, R. A., de Pinto, M. C., Valenti, D., Passarella, S., Marra, E., & De Gara, L. (2004). Production of reactive oxygen species, alteration of cytosolic ascorbate peroxidase, and impairment of mitochondrial metabolism are early events in heat shock-induced programmed cell death in tobacco Bright-Yellow 2 cells. *Plant physiology*, *134*(3), 1100-1112.

[82] Burbridge, E., Diamond, M., Dix, P. J., & McCabe, P. F. (2006). Use of cell morphology to evaluate the effect of a peroxidase gene on cell death induction thresholds in tobacco. *Plant Science*, *171*(1), 139-146.

[83] Zuppini, A., Andreoli, C., & Baldan, B. (2007). Heat stress: an inducer of programmed cell death in *Chlorella saccharophila*. *Plant and Cell Physiology*, *48*(7), 1000-1009.

[84] Curtis, M. J., & Wolpert, T. J. (2004). The victorin-induced mitochondrial permeability transition precedes cell shrinkage and biochemical markers of cell death, and shrinkage occurs without loss of membrane integrity. *The Plant Journal*, *38*(2), 244-259.

[85] Levine, A., Pennell, R. I., Alvarez, M. E., Palmer, R., & Lamb, C. (1996). Calcium-mediated apoptosis in a plant hypersensitive disease resistance response. *Current Biology*, *6*(4), 427-437.

[86] Ryerson, D. E., & Heath, M. C. (1996). Cleavage of nuclear DNA into oligonucleosomal fragments during cell death induced by fungal infection or by abiotic treatments. *The Plant Cell*, *8*(3), 393-402.

[87] Stein, J. C., & Hansen, G. (1999). Mannose induces an endonuclease responsible for DNA laddering in plant cells. *Plant Physiology*, *121*(1), 71-80.

[88] He, R., Drury, G. E., Rotari, V. I., Gordon, A., Willer, M., Farzaneh, T., ... & Gallois, P. (2008). Metacaspase-8 modulates programmed cell death induced by ultraviolet light and H2O2 in Arabidopsis. *Journal of Biological Chemistry*, *283*(2), 774-783.

[89] Buchanan-Wollaston, V., Page, T., Harrison, E., Breeze, E., Lim, P. O., Nam, H. G., ... & Leaver, C. J. (2005). Comparative transcriptome analysis reveals significant differences in gene expression and signalling pathways between developmental and dark/starvation-induced senescence in Arabidopsis. *The Plant Journal*, *42*(4), 567-585.

[90] De Jong, H. (2002). Modeling and simulation of genetic regulatory systems: a literature review. *Journal of computational biology*, *9*(1), 67-103.

[91] Kalra G., Bhatla S.C. (2018) Senescence and Programmed Cell Death. In: Plant Physiology, Development and Metabolism. Springer, Singapore. https://doi.org/10.1007/978-981-13-2023-1_30

[92] Yang, X., Liang, W., Chen, M., Zhang, D., Zhao, X. and Shi, J., 2017. Rice fatty acyl-CoA synthetase OsACOS12 is required for tapetum programmed cell death and male fertility. *Planta*, *246*(1), 105-122.

[93] Awwad, F., Bertrand, G., Grandbois, M. and Beaudoin, N., 2019. Auxin protects *Arabidopsis thaliana* cell suspension cultures from programmed cell death induced by the cellulose biosynthesis inhibitors thaxtomin A and

isoxaben. *BMC plant biology*, *19*(1), 1-13.

[94] Denbigh, G.L., Dauphinee, A.N., Fraser, M.S., Lacroix, C.R. and Gunawardena, A.H., 2020. The role of auxin in developmentally regulated programmed cell death in lace plant. *American journal of botany*, *107*(4), 577-586.

[95] Verma, N., Tiwari, S., Singh, V.P. and Prasad, S.M., 2020. Nitric oxide in plants: an ancient molecule with new tasks. *Plant Growth Regulation*, *90*(1), 1-13.

[96] Zhao, Y., Luo, L., Xu, J., Xin, P., Guo, H., Wu, J., Bai, L., Wang, G., Chu, J., Zuo, J. and Yu, H., 2018. Malate transported from chloroplast to mitochondrion triggers production of ROS and PCD in *Arabidopsis thaliana*. *Cell research*, *28*(4), 448-461.

[97] Poór, P., 2020. Effects of salicylic acid on the metabolism of mitochondrial reactive oxygen species in plants. *Biomolecules*, *10*(2), 341.

[98] Wang, H., Lin, J., Chang, Y. and Jiang, C.Z., 2017. Comparative transcriptomic analysis reveals that ethylene/H2O2-mediated hypersensitive response and programmed cell death determine the compatible interaction of sand pear and *Alternaria alternata*. *Frontiers in plant science*, *8*, 195.

[99] Ni, X.L., Gui, M.Y., Tan, L.L., Zhu, Q., Liu, W.Z. and Li, C.X., 2019. Programmed cell death and aerenchyma formation in water-logged sunflower stems and its promotion by ethylene and ROS. *Frontiers in plant science*, *9*, 1928.

[100] Hu, L., Zhou, K., Ren, G., Yang, S., Liu, Y., Zhang, Z., Li, Y., Gong, X. and Ma, F., 2020. Myo-inositol mediates reactive oxygen species-induced programmed cell death via salicylic acid-dependent and ethylene-dependent pathways in apple. *Horticulture research*, *7*(1), 1-13.

[101] Lu, Z., Sha, J., Tian, Y., Zhang, X., Liu, B. and Wu, Z., 2017. Polyphenolic allelochemical pyrogallic acid induces caspase-3 (like)-dependent programmed cell death in the cyanobacterium *Microcystis aeruginosa*. *Algal research*, *21*, 148-155.

[102] Araniti F, Costas-Gil A, Cabeiras-Freijanes L, Lupini A, Sunseri F, Reigosa MJ, Abenavoli MR, Sánchez-Moreiras AM. Rosmarinic acid induces programmed cell death in Arabidopsis seedlings through reactive oxygen species and mitochondrial dysfunction. PloS one. 2018, 26;13(12):e0208802.

[103] Mondal, R., & Das, P. (2020). Novel aspects of Cell Division Cycle and Apoptosis Regulator 1 (CCAR1) protein in *Morus notabilis*: an in silico approach. *Plant Signaling & Behavior*, *15*(10), 1795396.

Chapter 7

Regulation of Apoptosis during Environmental Skin Tumor Initiation

Bilal Bin Hafeez, Eunmi Park, Kyung-Soo Chun, Yong-Yeon Cho and Dae Joon Kim

Abstract

Skin cancer is more prevalent than any other cancer in the United States. Non-melanoma skin cancers are the more common forms of skin cancer that affect individuals. The development of squamous cell carcinoma, the second most common type of skin cancer, can be stimulated by exposure of environmental carcinogens, such as chemical toxicants or UVB. It is developed by three distinct stages: initiation, promotion, and progression. During the initiation, the fate of DNA-damaged skin cells is determined by the homeostatic regulation of pro-apoptotic and anti-apoptotic signaling pathways. The imbalance or disruption of either signaling will lead to the survival of initiated cells, resulting in the development of skin cancer. In this chapter, we will discuss signaling pathways that regulate apoptosis and the impact of their dysfunction during skin tumor initiation.

Keywords: apoptosis, tumor initiation, skin, carcinogen, UVB, AKT, STAT3, MAPK, TC-PTP

1. Introduction

Apoptosis, also called programmed cell death, is an essential cellular process that is required to maintain tissue homeostasis by modulating various biological functions including embryonic development, cellular growth, and immunity through the elimination of unwanted cells. Apoptosis accompanies the distinct morphological changes, such as membrane blebbing, cell shrinkage, chromatin condensation, and DNA fragmentation. These morphological changes during apoptosis were first observed by John Foxton R. Kerr, Andrew H. Wyllie, and Alastair Currie in 1972 with the first description of "apoptosis" [1]. Later, the critical role of programmed cell death occurred during the development of the nematode *Caenorhabditis elegans* was elucidated with the finding of death-related genes in 1999 [2]. Three scientists – Sydney Brenner, H. Robert Horvitz, and John E. Sulston – were awarded the Nobel Prize in Physiology or Medicine in 2002 for their discovery of genes that regulate apoptosis. Since then, apoptosis has extensively been studied as a crucial biochemical mechanism for the maintenance of normal tissue function and homeostasis. Dysregulation of apoptosis has been involved in various types of human diseases including cancers and neurodegenerative diseases [3, 4].

Cells can die prematurely by injury, infection, external assault such as chemicals, or a lack of blood supply through energy-independent process of cell death, which is called by necrosis. Necrosis is an unplanned cellular death by external damage. It causes severe disorganization and rupture of subcellular structure and eventually leads to disruption of the cell membrane. Then, the exposure of cellular components of necrotic cells into the surrounding interstitial tissue results in inflammation. In contrast to necrosis, apoptotic cell death is highly ordered and energy-dependent molecular process that can eliminate unwanted and damaged cells without causing inflammation [5, 6].

Apoptosis is initiated by two different pathways: the extrinsic and the intrinsic pathways. The extrinsic pathway is mediated by the interaction of transmembrane receptors known as death receptors with their ligands. Death receptors are members of the tumor necrosis factor (TNF) gene family and share similar cysteine-rich extracellular domains and homologous intracellular cytoplasmic sequence named as the 'death domain'. The death domain of receptors transmits the death signal from the cell surface to the intracellular signaling pathways to induce apoptosis in a rapid and efficient manner. For example, the binding of Fas ligand with Fas receptor, TNFα with TNF receptor, and TNF-related apoptosis-inducing ligand (TRAIL) with TRAIL receptor 1, 2 results in the interaction of the death domain of receptor with the adaptor protein, such as Fas-associated death domain (FADD) or TNF receptor-associated death domain (TRADD). The adaptor protein then associates with procaspase-8 and converts it into active caspase-8 [3, 7]. Likewise, the intrinsic pathway, which is known as the mitochondrial pathway, is activated by a broad range of exogenous and endogenous stimuli including radiation, toxins, viral infections, and oxidative stress. It modulates mitochondrial membrane potential through change of the ratio of pro-apoptotic and anti-apoptotic proteins located in the mitochondria and releases cytochrome c from the intermembrane space into the cytoplasm. Cytochrome c then forms a multi-protein complex called as the 'apoptosome' with apoptotic protease activating factor 1 (APAF1) and recruits inactive procaspase-9, which is followed by the dimerization and generation of catalytically active caspase-9 in an ATP-dependent manner [8, 9]. Initiation of both the extrinsic and intrinsic pathways can lead to the activation of executioner caspases, such as caspase-3, 6, or 7, resulting in the morphological changes of apoptosis [10]. These two apoptotic pathways are interconnected. The cleavage of pro-apoptotic protein Bid by caspase-8 can lead to its translocation to the mitochondria and result in the activation of the mitochondrial pathway [11]. Environment toxicants, such as ultraviolet B (UVB) radiation and chemical carcinogens, can cause severe DNA damage, which activates the intrinsic pathway in skin [12, 13]. UVB radiation can also activate the extrinsic pathway by inducing Fas receptor by p53 activation, promoting the association of TNF receptor 1 with TNF receptor-associated factor-2 (TRAF-2), or inhibiting the interaction of TRAIL receptors with decoy receptors that serve as their negative binding partners for apoptosis (**Figure 1**) [14–17].

In intrinsic pathway, mitochondrial outer membrane permeability is regulated by pro-apoptotic or anti-apoptotic proteins of B-cell lymphoma 2 (Bcl-2) family. A total of 25 genes in the Bcl-2 family have been identified. The pro-apoptotic proteins of Bcl-2 family include Bax, Bak, Bid, Bim, Bad, Noxa, Bmf, and Puma. The anti-apoptotic proteins of Bcl-2 family include Bcl-2, Bcl-X_L, Bcl-w, Mcl-1, and BAG. Altered expression of Bcl-2 family proteins has been involved in various diseases including cancer [18–21].

Skin, the largest organ of the human body, is continuously exposed to environmental assaults. Exposure of skin to the chemical carcinogens or UVB radiation induces a mutation in critical gene or genes and forms specific carcinogen-DNA adducts. These adducts can be repaired by DNA repair mechanism. Otherwise, they need to be removed by apoptosis. Defected apoptosis by genetic abnormalities and

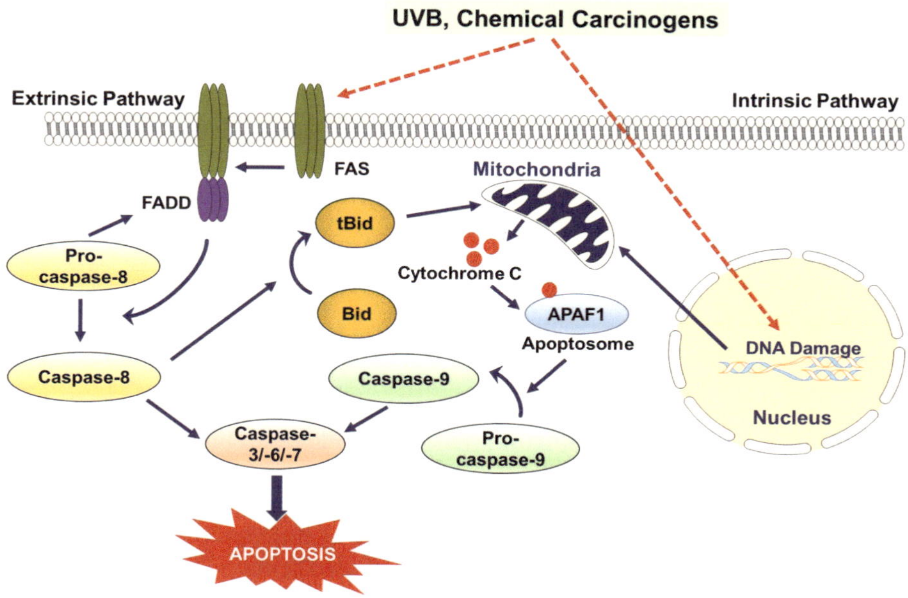

Figure 1.
Mechanisms of apoptosis induced by environmental carcinogens in skin. Apoptosis is initiated by two major pathways: The extrinsic and the intrinsic pathways. The extrinsic pathway is mediated by death receptors, such as Fas receptor. Upon binding to Fas ligand, Fas receptor interacts with FADD and recruits procaspase-8, which is followed by activation of caspase-8. The intrinsic pathway causes DNA damage and modulates mitochondrial membrane potential. Then, cytochrome c is released from the mitochondria into the cytoplasm. The formation of apoptosome with cytochrome c and APAF1 recruits and activates caspase-9. Activation of either caspase-8 or caspase-9 leads to the activation of executioner caspases including caspase-3, resulting in the morphological changes of apoptosis. These two pathways are also interconnected. The cleavage of bid by caspase-8 leads to its translocation to the mitochondria and activates the mitochondrial pathway. Environment toxicants, such as UVB radiation and chemical carcinogens, cause severe DNA damage and activate the intrinsic pathway in skin. UVB radiation activates the extrinsic pathway by activating death receptors and enhancing the association of death receptors with their ligands.

subsequent imbalanced signaling pathways can cause the survival of damaged cells during the tumor initiation. These damaged cells can undergo clonal expansion during the tumor promotion and eventually develop skin cancer [22–25]. In this chapter, we will review anti-apoptotic signaling pathways – AKT, STAT3, and MAPK – and pro-apoptotic signaling pathway – protein tyrosine phosphatases (PTPs) –, which are involved in the regulation of apoptosis during skin tumor initiation.

2. Anti-apoptotic signaling in skin tumor initiation

2.1 AKT signaling

AKT (protein kinase B) is the human homolog of the viral oncogene *v-akt* and is associated with protein kinases A (PKA) and C (PKC) in humans [26, 27]. The three known AKT isoforms are derived from distinct genes (AKT1/PKBα, AKT2/PKBβ and AKT3/PKBγ). The N terminus of AKT contains a pleckstrin homology domain, which can mediate lipid-protein and/or protein–protein interactions [28–30]. The C terminus of AKT contains a hydrophobic and proline-rich domain [28, 29]. AKT is activated by various growth factors such as platelet-derived and fibroblast growth factors and is involved in the regulation of cell survival signaling [31, 32]. AKT activation depends on its phosphorylation and four different phosphorylation sites on AKT (Ser-124, Thr-308, Thr-450, and Ser-473) have been identified [33]. Studies showed that extracellular stimuli induce the phosphorylation of AKT at Thr-308 and Ser-473 residues while the phosphorylation of AKT at Ser-124 and Thr-450 residues appears to be basally maintained [33]. Mutagenesis studies have revealed that the phosphorylation of AKT at Thr-308 and Ser-473 is required for its activation [34].

It has been shown that AKT signaling inactivates several pro-apoptotic factors including BAD, procaspase-9 and Forkhead transcription factors [35, 36]. In contrast, AKT activates various transcription factors that are involved in the upregulation of anti-apoptotic genes, such as cyclic-AMP response element-binding protein (CREB). It also activates IκB kinase (IKK) to phosphorylate IκB (inhibitor of NF-κB), leading to its proteasomal degradation and NF-κB nuclear localization. In addition, AKT reduces the protein levels of p53 by promoting its degradation through MDM2 phosphorylation, which can contribute to centrosome hyper-amplification and chromosome instability in cancer [37, 38]. Furthermore, AKT is involved in the regulation of subcellular localization of proteins. AKT can regulate the localization of various proteins and thereby their activity by phosphorylating specific binding sites for 14–3-3 proteins, which play a crucial role in the modulation of cellular location and degradation of proteins [39, 40].

Studies have shown that AKT is activated by environmental toxicants including UVB irradiation which protects keratinocytes against environmental attacks, implying its anti-apoptotic role in skin tumor initiation. It has been shown that UV radiation induces phosphorylation of AKT at Ser-473 and Thr-308 residues in mouse epidermal cell JB6 Cl41 in a time-dependent manner. These results were further confirmed by the observation that overexpression of AKT mutant, AKT-T308A/S473A, attenuated phosphorylation of AKT at Ser-473 and Thr-308 upon UVB irradiation [41]. The reactive oxygen species (ROS) generated by UV radiation acts as a mediator in UV-induced phosphorylation of AKT. It has been observed that pre-treatment of cells with either an antioxidant, N-acetyl-L-cysteine (NAC) or a specific antioxidant enzyme (catalase) inhibits phosphorylation of AKT in these cells, suggesting the link of ROS in UV radiation-induced activation of AKT [41]. Specific phosphorylation of Bad, a pro-apoptotic member of the Bcl-2 family, by AKT delayed the early activated apoptotic pathways in UVB-exposed human keratinocytes. AKT-mediated phosphorylation of Bad at serine 136 residue promoted its translocation from the mitochondria to the cytoplasm and subsequent cytoplasmic sequestration by 14–3-3ζ, resulting in the reduction of UVB-induced apoptosis in keratinocytes [42]. In addition to its anti-apoptotic function, studies using transgenic mice have shown that AKT can contribute to the development of skin cancer formation induced by environmental exposure through the regulation of epidermal proliferation. It has been shown that mice lacking $Akt1^{-/-};Akt2^{-/-}$ or $Akt1^{-/-};Akt3^{+/-}$ exhibit a hypoplastic epidermis due to decreased proliferation of keratinocytes while overexpression of wild type AKT1 (*wtAk*t) or constitutively active AKT (*myrAkt*) in the basal layer of mouse epidermis displays alterations in epidermal proliferation and differentiation [43–45]. Overexpression of either wtAKT or myrAKT in mouse epidermis displayed enhanced sensitivity to two stage skin carcinogenesis by promoting cell proliferation [45]. These studies suggest that AKT plays a critical role in the regulation of apoptosis and proliferation during skin carcinogenesis. Further detailed studies using transgenic mouse models will be helpful to elucidate the underlying molecular mechanisms and function of AKT in the initiation of skin carcinogenesis.

2.2 STAT3 signaling

Signal transducer and activator of transcription 3 (STAT3) is one of the family members of seven [STAT1 (α and β splice isoforms), STAT2 and STAT3 (α and β isoforms), STAT4, STAT5a, STAT5b, and STAT6)] latent cytoplasmic transcription factors which are encoded by seven individual genes [46]. STATs are phosphorylated at their specific tyrosine residue and activated by a wide variety of stimuli including growth factors and cytokines, which act through intrinsic receptor tyrosine kinases [47]. Tyrosine phosphorylation of STAT induces its dimerization via reciprocal interaction of phospho-tyrosine with Src homology domain 2 (SH2) between two STAT molecules. The phosphorylated STATs then translocate to the nucleus and bind to the consensus promoter sequences of downstream target genes, resulting in the activation of their transcription [48].

Different tyrosine kinases, such as RTKs (receptor tyrosine kinases) and non-RTKs including JAKs (Janus kinases), can phosphorylate STAT proteins. STAT tyrosine phosphorylation is transient, which lasts from 30 minutes to several hours in normal cells. However, studies have shown that STATs (specifically STAT3) are persistently tyrosine phosphorylated either as a consequence of deregulation of positive regulators of STAT activation such as tyrosine kinases or deactivation of negative regulators of STAT phosphorylation, such as phosphatases, suppressor of cytokine signaling, or protein inhibitor of activated STATs, in numerous cancer-derived cell lines or in primary tumors [49].

Studies have revealed that STAT3 is associated with cell survival and oncogenic transformation among the seven members of STATs. It has shown that targeted inhibition of STAT3 activation by antisense, small interfering RNA, dominant-negative STAT3 constructs, and/or blockade of tyrosine kinases has been associated with growth arrest and induction of apoptosis in cancer cell lines [49]. Furthermore, overexpression of constitutively activated STAT3 into immortalized cell lines led to oncogenic transformation, indicating a potential role of STAT3 in carcinogenesis [50, 51].

The generation of epidermal-specific STAT3-dificient and STAT3-overexpressing transgenic mice led to the main discovery of functional role of STAT3 in the initiation stage of skin carcinogenesis [52, 53]. Epidermal-specific deficiency of STAT3 in mouse *(K5Cre.Stat3*$^{fl/fl}$) significantly increased the sensitivity to apoptosis after 7,12-dimethylbenz[a]anthracene (DMBA) treatment both *in vitro* keratinocytes and *in vivo* epidermis compared with non-transgenic controls, as determined by increased caspase-3-positive cells. In particular, the significant increase in the number of keratinocyte stem cells (KSCs) undergoing apoptosis in the bulge region of hair follicles was observed in STAT3-deficient mice compared with non-transgenic littermates, indicating that STAT3 may be critical for maintaining the survival of KSCs during skin tumor initiation mediated by DMBA [54]. Similar with this observation, forced expression of a constitutively active form of STAT3 in mouse epidermis (*K5.Stat3C*) showed increased cell survival following DMBA exposure [55]. Further studies showed that STAT3 plays a critical role in the protection of damaged keratinocytes after UVB exposure. The epidermis of STAT3-deficient mice was highly sensitive to UVB-induced apoptosis compared with the epidermis of control mice, whereas the epidermis of K5.Stat3C mice was more resistant to UVB-induced apoptosis compared with the epidermis of control mice. UVB induces DNA damage and causes mutations in runs of tandemly located pyrimidine residues of DNA, resulting in the generation of cyclobutane pyrimidine dimers (CPDs) and pyrimidine (6–4) pyrimidone photoproducts (6–4 PPs). The number of either 6–4 PP- or CPD-positive cells was greater in epidermis of K5.Stat3C mice compared with epidermis from STAT3-deficient mice after UVB exposure [56]. Additionally, generation of inducible STAT3-deficient mice (*K5.Cre-ER*^{T2} *x Stat3*$^{fl/fl}$) clearly demonstrated that STAT3 has a critical role in the protection of keratinocytes during tumor initiation. Inducible deletion of STAT3 in mouse epidermis significantly increased apoptosis after DMBA treatment. Inducible STAT3 deletion in epidermis before initiation with DMBA showed a significant delay in tumor development and a significantly reduced number of tumors compared with control groups in two-stage skin carcinogenesis [57].

Expression of *Ha-ras homolog* (*v-Ha-ras*) into cultured primary keratinocytes *in vitro* has been utilized to generate initiated keratinocytes. Studies showed that the introduction of a STAT3 decoy molecule into the v-Ha-ras–initiated keratinocytes increases apoptosis with a concomitant decrease in Bcl-xL expression levels. In general, inhibition of STAT3 activation can lead to increase apoptosis or growth arrest in cancer-derived cell lines containing high levels of phosphorylated STAT3. However, STAT3 inhibition was not relatively affected in cell lines containing low or no levels of detectable tyrosine phosphorylated STAT3. Therefore, how STAT3 can protect keratinocytes against DMBA-induced apoptosis remains to be unclear, because DMBA does not induce tyrosine phosphorylation of STAT3 in keratinocytes, nor is it likely that v-Ha-ras–containing keratinocytes contain abundant levels of phosphorylated STAT3 [58]. One possible explanation is that the low levels of phosphorylated STAT3 present in these keratinocytes are sufficient to drive transcription of anti-apoptotic genes such as *Bcl-xL*. Alternatively, it is also possible that non-phosphorylated STAT3 may play a role as a transcription factor as previously demonstrated for STAT1 [59]. There are a few notable examples where relatively low levels of phosphorylated STAT3 are sufficient to mediate protection from growth arrest or apoptosis [60, 61]. Therefore, determination of the relative levels of phosphorylated STAT3 required to impart a phenotype is likely to be cell-type specific and remains an important objective.

It suggests that KSCs, which are located mostly within the bulge region of the hair follicle, are the major target cells for two-stage carcinogenesis [62]. The label-retaining cells (LRCs) retain the label for a sustained period of time following continuous administration of nucleotide analogs such as bromodeoxyuridine (BrdU) or [^{3}H] thymidine, indicating a very slow cycling frequency. Studies showed that hair follicle KSCs are identified within the LRCs [63, 64]. It has been observed that the STAT3-deficient keratinocytes undergoing apoptosis following DMBA exposure were located primarily within the bulge region of the hair follicle in an area adjacent to the LRC population. It indicates that the DMBA-sensitive cells may be KSCs, given their proximity to the LRCs. However, given the lack of overlap between the LRCs and the apoptotic cells, the cell type most sensitive to DMBA-induced apoptosis remains to be identified. Studies using inducible bulge-region KSC-specific STAT3-deficient mice (*K15.CrePR1 x Stat3*$^{fl/fl}$ mice) have provided further evidence that STAT3 is required for survival of bulge-region KSCs during the initiation stage of skin carcinogenesis [65]. In these studies, the number of apoptotic KSCs in the bulge-region was significantly increased in *K15.CrePR1 x Stat3*$^{fl/fl}$ mice by inducible deletion of STAT3 prior to tumor initiation with DMBA compared with control littermates. In addition, the frequency of *Ha-ras* codon 61 A182 → T mutations was decreased in *K15.CrePR1 x Stat3*$^{fl/fl}$ mice compared to control mice [65]. Furthermore, the number of skin tumors that developed in a two-stage skin carcinogenesis protocol was dramatically reduced by targeted deletion of STAT3 in bulge region KSCs at the time of initiation [65]. Overall, these studies provide molecular basis of STAT3 involvement in the initiation of skin carcinogenesis.

2.3 MAPK signaling

Mitogen-activated protein kinases (MAPKs) are essential signaling components that are vital in converting extracellular stimuli into intracellular responses through transcriptional regulation of various regulatory genes. MAPK signaling is activated by sequential protein kinase cascades including three enzymes: a MAPK kinase

kinase (MAPKKK), a MAPK kinase (MAPKK) and a MAPK [66]. MAPK signaling pathways are involved in the regulation of a wide variety of cellular processes including cell proliferation, differentiation, development, stress responses and apoptosis. Therefore, MAPK signaling is considered as one of potential therapeutic targets for many signaling-related diseases including cancer and diabetes. Three structurally related but biochemically and functionally distinct MAPKs are identified and named as extracellular signal-regulated kinases (ERKs), c-Jun N terminal kinases (JNKs) and p38 MAPKs [67–70].

ERK was the first MAPK identified and contains two isoforms: ERK1 and ERK2. ERK plays a critical role in the signaling of a variety of extracellular stimuli, such as growth factors and phorbol esters. ERK signaling is involved in the regulation of cell cycle progression and cell proliferation as one of major checkpoint signaling pathways for cellular mitogenesis [66, 71]. JNK was initially identified as a regulator of transcription factor c-Jun and consisted of three isoforms: JNK1, JNK2, and JNK3. JNK is also known as a stress-activated protein kinase (SAPK) as it is stimulated by various intra- or extracellular stresses. JNK signaling is involved in many cellular processes including immune response, neuronal activity, and insulin signaling [70, 72]. Studies have also shown that JNK signaling is critical in the promotion of apoptosis in response to a variety of harmful external stimuli through p53 activation [73, 74]. p38 MAPK contains four isoforms: p38α, p38β, p38δ, and p38γ. p38 MAPK is another SAPK and is activated by stress related stimuli. Similar with JNK, p38 MAPK is also involved in many cellular processes including apoptosis, inflammation, migration, differentiation, and cell cycle checkpoints [68, 75, 76].

In skin keratinocytes, both ERK and JNK are activated by UVB irradiation and protect cells against UVB-induced apoptosis [77, 78]. JNK is significantly activated by UVB exposure and pretreatment of antioxidant N-acetylcysteine reduced its activation, implying UVB-induced oxidative stress plays an important role in the activation of JNK [77]. Furthermore, UVB-mediated generation of reactive oxygen species significantly increased the activation of both JNK and ERK in human keratinocytes. Activated JNK and ERK then induced the upregulation of Bcl-2 and adenovirus E1B 19-kDa interacting protein 3 (BNIP3) expression, which is known to protect keratinocytes from UVB-induced apoptosis through autophagy. Pretreatment with the antioxidant N-acetylcysteine, the JNK inhibitor SP600125, or the ERK inhibitor U0126 significantly reduced the expression of BNIP3 upon UVB exposure and decreased cell survival by inducing apoptosis [78]. p38 MAPK is also activated by UVB irradiation [79]. However, in contrast to JNK and ERK, p38 MAPK promotes epidermal apoptosis following UVB exposure in mouse skin. The epidermis of transgenic mice that express a dominant negative p38α MAPK (p38DN) showed a significant reduction in UVB-induced apoptosis compared with the epidermis of control mice. The p38DN mice also showed a significant reduction of tumor number and growth compared to wild-type mice in UVB skin carcinogenesis assay [80]. Overall, it implies that JNK and ERK may protect damaged keratinocytes by reducing apoptosis during the initiation of skin carcinogenesis induced by environmental carcinogens, while p38 MAPK may contribute to remove damaged keratinocytes by promoting apoptosis in response to environmental exposure.

3. Pro-apoptotic signaling in skin tumor initiation

3.1 PTPs in skin

PTPs negatively regulate the rate and duration of phosphotyrosine signaling as an endogenous negative feedback mechanism of protein tyrosine kinases (PTKs)

[81–84]. PTPs were first identified in the late 1980s by Nicholas Tonks and colleagues 10 years after the discovery of PTKs [85]. Since then, 107 PTPs have been identified in the human genome by using the conserved catalytic domain of PTPs to search the human genome database [86, 87].

It has been shown that PTP expression is induced during proliferation and maturation of keratinocytes, however their expression levels remain unchanged within skin epidermis [88]. In contrast, microarray analysis has shown that expression of PTPs, such as PTPκ and PTPλ decreases in human melanoma tissue compared with its normal counterpart [89, 90]. Studies showed that acute ultraviolet (UV) exposure leads to the ligand-independent activation of PTKs [91, 92]. This result indicates that UV radiation may reduce PTP activation. With this regard, biochemical analysis showed that reactive oxygen species (such as H_2O_2) produced by UV irradiation caused the inactivation of PTPs through the oxidization of the cysteine residue located within the conserved active-site of the PTP catalytic domain [93–95]. Furthermore, studies performed by different groups have revealed that acute UV exposure can trigger PTP inactivation in keratinocytes [96, 97].

In contrast to previous findings of PTP inactivation observed in skin, studies revealed that three PTPs, T-cell protein tyrosine phosphatase (TC-PTP), Src homology region 2 domain-containing phosphatase 1 (SHP1), and SHP2, can cooperate in the dephosphorylation of STAT3 in response to UVB irradiation [98]. STAT3 was rapidly dephosphorylated in keratinocytes after UVB irradiation. Knockdown of TC-PTP, SHP1, or SHP2 using RNAi in keratinocytes before UVB exposure partially recovered the level of phosphorylated STAT3 at Tyr 705 (PY-STAT3) compared to control keratinocytes, indicating that these PTPs are responsible for the rapid STAT3 dephosphorylation observed following UVB exposure. Further studies revealed that knockdown of all three phosphatases, using RNAi, prevented the rapid dephosphorylation of STAT3 following UVB irradiation [98]. This result suggests that exposure to UVB triggers PTP activation, which attenuates STAT3 signaling by dephosphorylating PY-STAT3. It implies that this activation of PTP can contribute to increase UVB-induced apoptosis during tumor initiation by deactivating STAT3, one of major survival factors in skin.

3.2 TC-PTP/PTPN2 signaling

Among three PTPs involved in STAT3 dephosphorylation of keratinocytes after UVB exposure, further investigation revealed that TC-PTP is the major PTP involved in the regulation of STAT3 signaling in keratinocytes following UVB exposure. TC-PTP activity was steadily increased after treatment of low-dose of UVB (10 mJ/cm^2), which can contribute to STAT3 dephosphorylation in mouse keratinocytes. Knockdown of TC-PTP in mouse 3PC keratinocytes significantly suppressed UVB-induced apoptosis with decreased caspase-3 activity compared to control keratinocytes [99].

TC-PTP was one of the first members of the PTP gene family to be identified. It is encoded by protein tyrosine phosphatase non-receptor type 2 (PTPN2). As one of 17 intracellular, non-receptor PTPs, TC-PTP is broadly expressed in most embryonic and adult tissues, but it is highly expressed in hematopoietic tissues [100, 101]. Two different forms of TC-PTP are generated by alternative splicing at the 3′ end of the gene: TC45 (TC-PTPa) and TC48 (TC-PTPb). TC45 (45 kDa) is the major form of TC-PTP in most species, including humans and mice. TC45 is mainly localized in the nucleus with a bipartite nuclear localization signal (NLS) in its C-terminal domain, while TC48 (48 kDa) is localized to the endoplasmic reticulum with its hydrophobic C terminus. Studies have shown that almost all TC-PTP mRNA encodes TC45 in mouse tissue and TC48 mRNA is not detectable by Northern blot analysis [102–104].

Recent generation of epidermal-specific TC-PTP knockout (*K14Cre.Ptpn2*$^{fl/fl}$; TC-PTP KO) transgenic mice as *in vivo* models has provided an evidence that TC-PTP plays a crucial role in the promotion of epidermal apoptosis induced by environmental assaults [105, 106]. TC-PTP deficiency in mouse epidermis led to a desensitization to tumor initiator DMBA-induced apoptosis both *in vivo* epidermis and *in vitro* keratinocytes. The number of apoptotic cells, detected by active caspase-3 staining, within the epidermis of control (TC-PTP WT) mice was significantly increased compared to TC-PTP KO mouse epidermis following DMBA treatment. Profound morphological changes induced by apoptosis, such as cell ballooning and bleb formation, were found in TC-PTP WT keratinocytes compared to TC-PTP KO keratinocytes. Similarly, annexin V-positive cells and caspase-3 activity were significantly increased in TC-PTP WT keratinocytes compared to TC-PTP KO keratinocytes. Inhibition of STAT3 or AKT in TC-PTP KO keratinocytes significantly reversed the effects of TC-PTP deficiency on apoptosis by increasing cellular sensitivity and caspase-3 activity following DMBA treatment compared to control keratinocytes [105]. Further studies also showed that TC-PTP KO cells showed increased survival against UVB-induced apoptosis compared to control, which was concomitant with a UVB-mediated increase in the level of Flk-1 (fetal liver kinase-1, known as VEGFR2) phosphorylation. Immunoprecipitation of the TC-PTP substrate-trapping mutant TCPTP-D182A showed that TC-PTP directly interacts with Flk-1 to dephosphorylate it and their interaction was stimulated by UVB irradiation. Following UVB-mediated Flk-1 phosphorylation in the absence of TC-PTP, the level of phosphorylated JNK was significantly increased in TC-PTP KO cells compared to TC-PTP WT cells after UVB irradiation. Inhibition of Flk-1 or JNK by their specific inhibitors in TC-PTP KO cells reversed this effect and significantly increased UVB-induced apoptosis compared to untreated TC-PTP KO cells [106].

The nuclear form of TC-PTP (TC45) contains bipartite nuclear localization signals (NLSI and NLSII) in its C-terminus, and TC45 had been known to be primarily localized in the cell nucleus due to its NLS [103]. However, recent studies have showed that TC45 is mainly localized to the cytoplasm of keratinocytes. TC45 is translocated to the nucleus following UVB irradiation [98]. TC45 nuclear translocation increased its activity in the nucleus and resulted in an increase of UVB-induced apoptosis which corresponded to a decrease in nuclear phosphorylated STAT3. UVB irradiation activated AKT to trigger nuclear translocation of TC45 and the adaptor protein 14–3-3σ. Furthermore, site-directed mutagenesis of putative 14–3-3σ binding sites within TC45 revealed that a substitution at threonine 179 (TC45/T179A) effectively blocked UVB-induced nuclear translocation of exogenous TC45 due to the disruption of the direct binding between TC45 and 14–3-3σ. Overexpression of TC45/T179A in keratinocytes resulted in decreased UVB-induced apoptosis, indicating that TC45 nuclear translocation is an important step to induce apoptosis against UVB-mediated damage [107].

Recent generation of epidermal-specific TC-PTP-overexpressing (*K5HA.Ptpn2*) mouse model further demonstrated that TC-PTP has a critical role for the induction of epidermal apoptosis against chemical carcinogen [108]. Overexpression of TC-PTP increased epidermal sensitivity to DMBA-induced apoptosis through the synergistic regulation of STAT1, STAT3, STAT5, and PI3K/AKT signaling. Inhibition of STAT1, STAT3, STAT5, or AKT reversed the effects of TC-PTP overexpression on epidermal apoptosis after DMBA treatment [108].

Overall, these studies suggest that TC-PTP plays a protective role against environmental carcinogens by increasing epidermal apoptosis through the regulation of AKT, STAT including STAT3, and Flk-1/JNK signaling pathways (**Figure 2**).

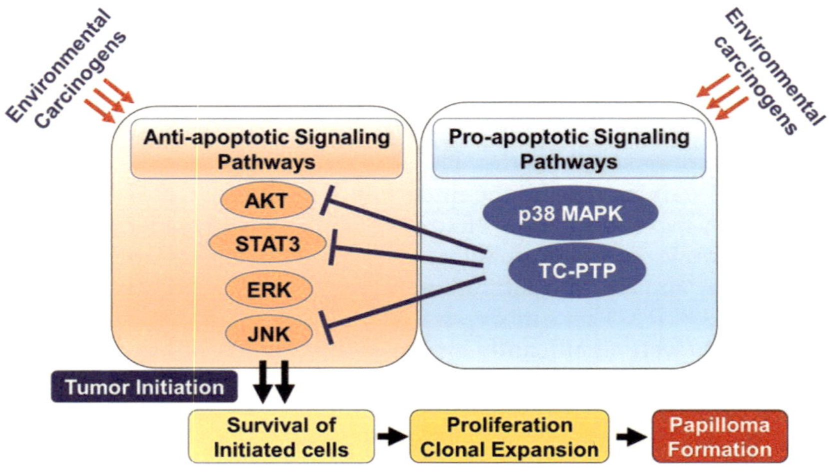

Figure 2.
Anti-apoptotic and pro-apoptotic signaling pathways in the regulation of skin tumor initiation induced by environmental carcinogens. AKT, STAT3, ERK, and JNK signaling pathways inhibit epidermal apoptosis in response to environmental carcinogens, such as UVB and chemical carcinogens/toxins, which can lead to survival of DNA damaged cells during tumor initiation. These initiated cells are clonally expanded by accelerated proliferation during tumor promotion and form benign skin tumor called by 'papilloma'. On the other hand, TC-PTP signaling pathway is initially activated against environmental exposure. TC-PTP signaling attenuates AKT, STAT3, and JNK signaling pathway through either direct or indirect dephosphorylation and increases epidermal apoptosis as one of initial protective mechanisms in skin. p38 MAPK signaling pathway is also activated and may contribute to increase epidermal apoptosis following environmental exposure.

4. Conclusion

Apoptosis is a critical cellular process to protect keratinocytes during skin tumor initiation. Environmental assaults induce a mutation in a critical gene or genes. Apoptosis can remove DNA-damaged keratinocytes that are not repaired via DNA damage repair mechanism. Survival of damaged keratinocytes by increased anti-apoptotic signaling pathways or decreased pro-apoptotic signaling pathways will lead to clonal expansion of them during skin tumor promotion, which is followed by the formation of papilloma and then squamous cell carcinoma. In this chapter, we outline the functional significance of three – AKT, STAT3, and MAPK – anti-apoptotic signaling pathways and TC-PTP as a pro-apoptotic signaling pathway in the regulation of apoptosis during skin tumor initiation. Besides them, different signaling pathways are involved in modulating apoptosis during skin tumor initiation depending on the types of environmental assaults and interconnected with signaling pathways we mentioned in this chapter. Further understanding of signaling mechanisms and their function in environmentally induced epidermal apoptosis during tumor initiation will contribute to develop novel therapeutic interventions for the prevention and treatment of skin cancer.

Acknowledgements

This work was supported by the institutional fund from University of Texas Rio Grande Valley (to D.J. Kim) and the Ministry of Science, ICT and Future Planning (NRF-2020R1A2B5B02001804) (to Y.-Y. Cho).

Conflict of interest

No potential conflicts of interest were disclosed.

Competing financial interests

The authors declare no competing financial interests.

Appendices and nomenclature

6-4PP	pyrimidine (6–4) pyrimidone photoproduct
AKT	a serine/threonine specific protein kinase, known as protein kinase B (PKB)
APAF1	apoptotic protease activating factor 1
Bcl-2	B-cell lymphoma 2
BNIP3	Bcl-2 and adenovirus E1B 19-kDa interacting protein 3
Caspase	cysteine-aspartic protease, cysteine aspartase or cysteine-dependent aspartate-directed protease
CREB	cyclic-AMP response element-binding protein
CPD	cyclobutane pyrimidine dimer
DMBA	7,12-dimethylbenz[a]anthracene
ERK	extracellular signal-regulated kinase
FADD	Fas-associated death domain
Flk-1	fetal liver kinase-1, known as VEGFR2
JAK	Janus kinase
JNK	c-Jun N-terminal kinase
K5	keratin 5
K14	keratin 14
KO	knockout
KSC	keratinocyte stem cell
LRC	label-retaining cell
MAPK	mitogen-activated protein kinase
NLS	nuclear localization signal
p38DN	dominant negative p38α MAPK
PTP	protein tyrosine phosphatase
PTK	protein tyrosine kinase
PTPN2	protein tyrosine phosphatase non-receptor type 2
PY-STAT3	phosphorylated STAT3 on the tyrosine residue
RNAi	RNA interference
RTK	receptor tyrosine kinase
SAPK	stress-activated protein kinase
SHP-1	Src homology region 2 domain-containing phosphatase-1
SHP-2	Src homology region 2 domain-containing phosphatase-2
STAT3	signal transducer and activator of transcription 3
TC-PTP	T-cell protein tyrosine phosphatase
TRADD	TNF receptor-associated death domain
TRAF-2	TNF receptor-associated factor-2
TRAIL	TNF-related apoptosis-inducing ligand
TNF	tumor necrosis factor
UVB	ultraviolet B
WT	wild type

Author details

Bilal Bin Hafeez[1], Eunmi Park[2], Kyung-Soo Chun[3], Yong-Yeon Cho[4] and Dae Joon Kim[5,6]*

1 Department of Microbiology and Immunology, School of Medicine, University of Texas Rio Grande Valley, Edinburg, TX, USA

2 Department of Food and Nutrition, Hannam University, Daejeon, Republic of Korea

3 College of Pharmacy, Keimyung University, Daegu, Republic of Korea

4 BRL and BK21-4th team, College of Pharmacy, The Catholic University of Korea, Bucheon-si, Republic of Korea

5 Department of Molecular Science, School of Medicine, University of Texas Rio Grande Valley, Edinburg, TX, USA

6 Department of Human Genetics, School of Medicine, University of Texas Rio Grande Valley, Edinburg, TX, USA

*Address all correspondence to: dae.kim@utrgv.edu

References

[1] Kerr JF, Wyllie AH, Currie AR. Apoptosis: a basic biological phenomenon with wide-ranging implications in tissue kinetics. Br J Cancer 1972; 26: 239-257.

[2] Horvitz HR. Genetic control of programmed cell death in the nematode *Caenorhabditis elegans*. Cancer Res 1999; 59: 1701s–.

[3] Elmore S. Apoptosis: a review of programmed cell death. Toxicol Pathol 2007; 35: 495-516.

[4] Wong RS. Apoptosis in cancer: from pathogenesis to treatment. J Exp Clin Cancer Res 2011; 30: 87.

[5] Kanduc D, Mittelman A, Serpico R, Sinigaglia E, Sinha AA, Natale C *et al*. Cell death: apoptosis versus necrosis (review). Int J Oncol 2002; 21: 165-170.

[6] Nicotera P, Melino G. Regulation of the apoptosis-necrosis switch. Oncogene 2004; 23: 2757-2765.

[7] Kumar R, Herbert PE, Warrens AN. An introduction to death receptors in apoptosis. Int J Surg 2005; 3: 268-277.

[8] Riedl SJ, Salvesen GS. The apoptosome: signalling platform of cell death. Nat Rev Mol Cell Biol 2007; 8: 405-413.

[9] Bao Q, Shi Y. Apoptosome: a platform for the activation of initiator caspases. Cell Death Differ 2007; 14: 56-65.

[10] Cohen GM. Caspases: the executioners of apoptosis. Biochem J 1997; 326 (Pt 1): 1-16.

[11] Li H, Zhu H, Xu CJ, Yuan J. Cleavage of BID by caspase 8 mediates the mitochondrial damage in the Fas pathway of apoptosis. Cell 1998; 94: 491-501.

[12] Lee CH, Wu SB, Hong CH, Yu HS, Wei YH. Molecular Mechanisms of UV-Induced Apoptosis and Its Effects on Skin Residential Cells: The Implication in UV-Based Phototherapy. Int J Mol Sci 2013; 14: 6414-6435.

[13] Raj D, Brash DE, Grossman D. Keratinocyte apoptosis in epidermal development and disease. J Invest Dermatol 2006; 126: 243-257.

[14] Bang B, Gniadecki R, Larsen JK, Baadsgaard O, Skov L. In vivo UVB irradiation induces clustering of Fas (CD95) on human epidermal cells. Exp Dermatol 2003; 12: 791-798.

[15] Aragane Y, Kulms D, Metze D, Wilkes G, Poppelmann B, Luger TA *et al*. Ultraviolet light induces apoptosis via direct activation of CD95 (Fas/APO-1) independently of its ligand CD95L. J Cell Biol 1998; 140: 171-182.

[16] Tobin D, van Hogerlinden M, Toftgard R. UVB-induced association of tumor necrosis factor (TNF) receptor 1/TNF receptor-associated factor-2 mediates activation of Rel proteins. Proc Natl Acad Sci U S A 1998; 95: 565-569.

[17] Qin JZ, Bacon P, Panella J, Sitailo LA, Denning MF, Nickoloff BJ. Low-dose UV-radiation sensitizes keratinocytes to TRAIL-induced apoptosis. J Cell Physiol 2004; 200: 155-166.

[18] Yip KW, Reed JC. Bcl-2 family proteins and cancer. Oncogene 2008; 27: 6398-6406.

[19] Brunelle JK, Letai A. Control of mitochondrial apoptosis by the Bcl-2 family. J Cell Sci 2009; 122: 437-441.

[20] Czabotar PE, Lessene G, Strasser A, Adams JM. Control of apoptosis by the BCL-2 protein family: implications for physiology and therapy. Nat Rev Mol Cell Biol 2014; 15: 49-63.

[21] Roy MJ, Vom A, Czabotar PE, Lessene G. Cell death and the mitochondria: therapeutic targeting of the BCL-2 family-driven pathway. Br J Pharmacol 2014; 171: 1973-1987.

[22] DiGiovanni J. Multistage carcinogenesis in mouse skin. Pharmacol Ther 1992; 54: 63-128.

[23] Zoumpourlis V, Solakidi S, Papathoma A, Papaevangeliou D. Alterations in signal transduction pathways implicated in tumour progression during multistage mouse skin carcinogenesis. Carcinogenesis 2003; 24: 1159-1165.

[24] de Gruijl FR, van Kranen HJ, Mullenders LH. UV-induced DNA damage, repair, mutations and oncogenic pathways in skin cancer. J Photochem Photobiol B 2001; 63: 19-27.

[25] Melnikova VO, Ananthaswamy HN. Cellular and molecular events leading to the development of skin cancer. Mutat Res 2005; 571: 91-106.

[26] Coffer PJ, Woodgett JR. Molecular cloning and characterisation of a novel putative protein-serine kinase related to the cAMP-dependent and protein kinase C families. Eur J Biochem 1991; 201: 475-481.

[27] Bellacosa A, Testa JR, Staal SP, Tsichlis PN. A retroviral oncogene, akt, encoding a serine-threonine kinase containing an SH2-like region. Science 1991; 254: 274-277.

[28] Datta SR, Brunet A, Greenberg ME. Cellular survival: a play in three Akts. Genes Dev 1999; 13: 2905-2927.

[29] Mayer BJ, Ren R, Clark KL, Baltimore D. A putative modular domain present in diverse signaling proteins. Cell 1993; 73: 629-630.

[30] Musacchio A, Gibson T, Rice P, Thompson J, Saraste M. The PH domain: a common piece in the structural patchwork of signalling proteins. Trends Biochem Sci 1993; 18: 343-348.

[31] Burgering BM, Coffer PJ. Protein kinase B (c-Akt) in phosphatidylinositol-3-OH kinase signal transduction. Nature 1995; 376: 599-602.

[32] Franke TF, Yang SI, Chan TO, Datta K, Kazlauskas A, Morrison DK *et al*. The protein kinase encoded by the Akt proto-oncogene is a target of the PDGF-activated phosphatidylinositol 3-kinase. Cell 1995; 81: 727-736.

[33] Alessi DR, Andjelkovic M, Caudwell B, Cron P, Morrice N, Cohen P *et al*. Mechanism of activation of protein kinase B by insulin and IGF-1. EMBO J 1996; 15: 6541-6551.

[34] Jiang BH, Aoki M, Zheng JZ, Li J, Vogt PK. Myogenic signaling of phosphatidylinositol 3-kinase requires the serine-threonine kinase Akt/protein kinase B. Proc Natl Acad Sci U S A 1999; 96: 2077-2081.

[35] Franke TF. PI3K/Akt: getting it right matters. Oncogene 2008; 27: 6473-6488.

[36] Franke TF, Hornik CP, Segev L, Shostak GA, Sugimoto C. PI3K/Akt and apoptosis: size matters. Oncogene 2003; 22: 8983-8998.

[37] Carroll PE, Okuda M, Horn HF, Biddinger P, Stambrook PJ, Gleich LL *et al*. Centrosome hyperamplification in human cancer: chromosome instability induced by p53 mutation and/or Mdm2 overexpression. Oncogene 1999; 18: 1935-1944.

[38] Sherr CJ, Weber JD. The ARF/p53 pathway. Curr Opin Genet Dev 2000; 10: 94-99.

[39] Cahill CM, Tzivion G, Nasrin N, Ogg S, Dore J, Ruvkun G *et al*. Phosphatidylinositol 3-kinase signaling inhibits DAF-16 DNA binding and

function via 14-3-3-dependent and 14-3-3-independent pathways. J Biol Chem 2001; 276: 13402-13410.

[40] Brunet A, Bonni A, Zigmond MJ, Lin MZ, Juo P, Hu LS *et al*. Akt promotes cell survival by phosphorylating and inhibiting a Forkhead transcription factor. Cell 1999; 96: 857-868.

[41] Huang C, Li J, Ding M, Leonard SS, Wang L, Castranova V *et al*. UV Induces phosphorylation of protein kinase B (Akt) at Ser-473 and Thr-308 in mouse epidermal Cl 41 cells through hydrogen peroxide. J Biol Chem 2001; 276: 40234-40240.

[42] Claerhout S, Decraene D, Van Laethem A, Van Kelst S, Agostinis P, Garmyn M. AKT delays the early-activated apoptotic pathway in UVB-irradiated keratinocytes via BAD translocation. J Invest Dermatol 2007; 127: 429-438.

[43] Peng XD, Xu PZ, Chen ML, Hahn-Windgassen A, Skeen J, Jacobs J *et al*. Dwarfism, impaired skin development, skeletal muscle atrophy, delayed bone development, and impeded adipogenesis in mice lacking Akt1 and Akt2. Genes Dev 2003; 17: 1352-1365.

[44] Yang ZZ, Tschopp O, Di-Poi N, Bruder E, Baudry A, Dummler B *et al*. Dosage-dependent effects of Akt1/protein kinase Balpha (PKBalpha) and Akt3/PKBgamma on thymus, skin, and cardiovascular and nervous system development in mice. Mol Cell Biol 2005; 25: 10407-10418.

[45] Segrelles C, Lu J, Hammann B, Santos M, Moral M, Cascallana JL *et al*. Deregulated activity of Akt in epithelial basal cells induces spontaneous tumors and heightened sensitivity to skin carcinogenesis. Cancer research 2007; 67: 10879-10888.

[46] Klampfer L. Signal transducers and activators of transcription (STATs): Novel targets of chemopreventive and chemotherapeutic drugs. Curr Cancer Drug Targets 2006; 6: 107-121.

[47] Hodge DR, Hurt EM, Farrar WL. The role of IL-6 and STAT3 in inflammation and cancer. Eur J Cancer 2005; 41: 2502-2512.

[48] Darnell JE, Jr. STATs and gene regulation. Science 1997; 277: 1630-1635.

[49] Bromberg J. Stat proteins and oncogenesis. J Clin Invest 2002; 109: 1139-1142.

[50] Dechow TN, Pedranzini L, Leitch A, Leslie K, Gerald WL, Linkov I *et al*. Requirement of matrix metalloproteinase-9 for the transformation of human mammary epithelial cells by Stat3-C. Proc Natl Acad Sci U S A 2004; 101: 10602-10607.

[51] Bromberg JF, Wrzeszczynska MH, Devgan G, Zhao Y, Pestell RG, Albanese C *et al*. Stat3 as an oncogene. Cell 1999; 98: 295-303.

[52] Sano S, Chan KS, DiGiovanni J. Impact of Stat3 activation upon skin biology: a dichotomy of its role between homeostasis and diseases. J Dermatol Sci 2008; 50: 1-14.

[53] Kim DJ, Chan KS, Sano S, Digiovanni J. Signal transducer and activator of transcription 3 (Stat3) in epithelial carcinogenesis. Mol Carcinog 2007; 46: 725-731.

[54] Chan KS, Sano S, Kiguchi K, Anders J, Komazawa N, Takeda J *et al*. Disruption of Stat3 reveals a critical role in both the initiation and the promotion stages of epithelial carcinogenesis. J Clin Invest 2004; 114: 720-728.

[55] Chan KS, Sano S, Kataoka K, Abel E, Carbajal S, Beltran L *et al*. Forced expression of a constitutively active form of Stat3 in mouse epidermis enhances

malignant progression of skin tumors induced by two-stage carcinogenesis. Oncogene 2008; 27: 1087-1094.

[56] Kim DJ, Angel JM, Sano S, DiGiovanni J. Constitutive activation and targeted disruption of signal transducer and activator of transcription 3 (Stat3) in mouse epidermis reveal its critical role in UVB-induced skin carcinogenesis. Oncogene 2009; 28: 950-960.

[57] Kataoka K, Kim DJ, Carbajal S, Clifford JL, DiGiovanni J. Stage-specific disruption of Stat3 demonstrates a direct requirement during both the initiation and promotion stages of mouse skin tumorigenesis. Carcinogenesis 2008; 29: 1108-1114.

[58] Turkson J, Bowman T, Garcia R, Caldenhoven E, De Groot RP, Jove R. Stat3 activation by Src induces specific gene regulation and is required for cell transformation. Mol Cell Biol 1998; 18: 2545-2552.

[59] Chatterjee-Kishore M, Wright KL, Ting JP, Stark GR. How Stat1 mediates constitutive gene expression: a complex of unphosphorylated Stat1 and IRF1 supports transcription of the LMP2 gene. EMBO J 2000; 19: 4111-4122.

[60] Niu G, Bowman T, Huang M, Shivers S, Reintgen D, Daud A *et al*. Roles of activated Src and Stat3 signaling in melanoma tumor cell growth. Oncogene 2002; 21: 7001-7010.

[61] Li L, Shaw PE. Autocrine-mediated activation of STAT3 correlates with cell proliferation in breast carcinoma lines. J Biol Chem 2002; 277: 17397-17405.

[62] Morris RJ. Keratinocyte stem cells: targets for cutaneous carcinogens. J Clin Invest 2000; 106: 3-8.

[63] Morris RJ, Liu Y, Marles L, Yang Z, Trempus C, Li S *et al*. Capturing and profiling adult hair follicle stem cells. Nat Biotechnol 2004; 22: 411-417.

[64] Tumbar T, Guasch G, Greco V, Blanpain C, Lowry WE, Rendl M *et al*. Defining the epithelial stem cell niche in skin. Science 2004; 303: 359-363.

[65] Kim DJ, Kataoka K, Rao D, Kiguchi K, Cotsarelis G, Digiovanni J. Targeted disruption of stat3 reveals a major role for follicular stem cells in skin tumor initiation. Cancer Res 2009; 69: 7587-7594.

[66] Zhang W, Liu HT. MAPK signal pathways in the regulation of cell proliferation in mammalian cells. Cell Res 2002; 12: 9-18.

[67] Braicu C, Buse M, Busuioc C, Drula R, Gulei D, Raduly L *et al*. A Comprehensive Review on MAPK: A Promising Therapeutic Target in Cancer. Cancers (Basel) 2019; 11.

[68] Plotnikov A, Zehorai E, Procaccia S, Seger R. The MAPK cascades: signaling components, nuclear roles and mechanisms of nuclear translocation. Biochim Biophys Acta 2011; 1813: 1619-1633.

[69] Cargnello M, Roux PP. Activation and function of the MAPKs and their substrates, the MAPK-activated protein kinases. Microbiol Mol Biol Rev 2011; 75: 50-83.

[70] Kyriakis JM, Avruch J. Mammalian MAPK signal transduction pathways activated by stress and inflammation: a 10-year update. Physiol Rev 2012; 92: 689-737.

[71] Dhillon AS, Hagan S, Rath O, Kolch W. MAP kinase signalling pathways in cancer. Oncogene 2007; 26: 3279-3290.

[72] Johnson GL, Nakamura K. The c-jun kinase/stress-activated pathway:

regulation, function and role in human disease. Biochim Biophys Acta 2007; 1773: 1341-1348.

[73] Fuchs SY, Adler V, Pincus MR, Ronai Z. MEKK1/JNK signaling stabilizes and activates p53. Proc Natl Acad Sci U S A 1998; 95: 10541-10546.

[74] Zeke A, Misheva M, Remenyi A, Bogoyevitch MA. JNK Signaling: Regulation and Functions Based on Complex Protein-Protein Partnerships. Microbiol Mol Biol Rev 2016; 80: 793-835.

[75] Cuadrado A, Nebreda AR. Mechanisms and functions of p38 MAPK signalling. Biochem J 2010; 429: 403-417.

[76] Cuenda A, Rousseau S. p38 MAP-kinases pathway regulation, function and role in human diseases. Biochim Biophys Acta 2007; 1773: 1358-1375.

[77] Assefa Z, Garmyn M, Bouillon R, Merlevede W, Vandenheede JR, Agostinis P. Differential stimulation of ERK and JNK activities by ultraviolet B irradiation and epidermal growth factor in human keratinocytes. J Invest Dermatol 1997; 108: 886-891.

[78] Moriyama M, Moriyama H, Uda J, Kubo H, Nakajima Y, Goto A *et al*. BNIP3 upregulation via stimulation of ERK and JNK activity is required for the protection of keratinocytes from UVB-induced apoptosis. Cell Death Dis 2017; 8: e2576.

[79] Peus D, Vasa RA, Beyerle A, Meves A, Krautmacher C, Pittelkow MR. UVB activates ERK1/2 and p38 signaling pathways via reactive oxygen species in cultured keratinocytes. J Invest Dermatol 1999; 112: 751-756.

[80] Dickinson SE, Olson ER, Zhang J, Cooper SJ, Melton T, Criswell PJ *et al*. p38 MAP kinase plays a functional role in UVB-induced mouse skin carcinogenesis. Mol Carcinog 2011; 50: 469-478.

[81] Lemmon MA, Schlessinger J. Cell signaling by receptor tyrosine kinases. Cell 2010; 141: 1117-1134.

[82] Tonks NK, Neel BG. From form to function: signaling by protein tyrosine phosphatases. Cell 1996; 87: 365-368.

[83] Stoker AW. Protein tyrosine phosphatases and signalling. J Endocrinol 2005; 185: 19-33.

[84] Hendriks WJ, Elson A, Harroch S, Pulido R, Stoker A, den Hertog J. Protein tyrosine phosphatases in health and disease. FEBS J 2012.

[85] Tonks NK, Diltz CD, Fischer EH. Purification of the major protein-tyrosine-phosphatases of human placenta. J Biol Chem 1988; 263: 6722-6730.

[86] Alonso A, Sasin J, Bottini N, Friedberg I, Friedberg I, Osterman A *et al*. Protein tyrosine phosphatases in the human genome. Cell 2004; 117: 699-711.

[87] Julien SG, Dube N, Hardy S, Tremblay ML. Inside the human cancer tyrosine phosphatome. Nat Rev Cancer 2011; 11: 35-49.

[88] Hendriks W, Brugman C, Richter KH, van Hooijdonk C, Schepens J, Schalkwijk J *et al*. Protein-tyrosine phosphatases expressed in mouse epidermal keratinocytes. J Invest Dermatol 1996; 106: 972-976.

[89] McArdle L, Rafferty MM, Satyamoorthy K, Maelandsmo GM, Dervan PA, Herlyn M *et al*. Microarray analysis of phosphatase gene expression in human melanoma. Br J Dermatol 2005; 152: 925-930.

[90] McArdle L, Rafferty M, Maelandsmo GM, Bergin O, Farr CJ, Dervan PA *et al*. Protein tyrosine phosphatase genes downregulated in melanoma. J Invest Dermatol 2001; 117: 1255-1260.

[91] Sachsenmaier C, Radler-Pohl A, Zinck R, Nordheim A, Herrlich P, Rahmsdorf HJ. Involvement of growth factor receptors in the mammalian UVC response. Cell 1994; 78: 963-972.

[92] Coffer PJ, Burgering BM, Peppelenbosch MP, Bos JL, Kruijer W. UV activation of receptor tyrosine kinase activity. Oncogene 1995; 11: 561-569.

[93] Caselli A, Marzocchini R, Camici G, Manao G, Moneti G, Pieraccini G *et al*. The inactivation mechanism of low molecular weight phosphotyrosine-protein phosphatase by H2O2. J Biol Chem 1998; 273: 32554-32560.

[94] Denu JM, Tanner KG. Redox regulation of protein tyrosine phosphatases by hydrogen peroxide: detecting sulfenic acid intermediates and examining reversible inactivation. Methods Enzymol 2002; 348: 297-305.

[95] Tonks NK. Redox redux: revisiting PTPs and the control of cell signaling. Cell 2005; 121: 667-670.

[96] Gulati P, Markova B, Gottlicher M, Bohmer FD, Herrlich PA. UVA inactivates protein tyrosine phosphatases by calpain-mediated degradation. EMBO Rep 2004; 5: 812-817.

[97] Xu Y, Shao Y, Voorhees JJ, Fisher GJ. Oxidative inhibition of receptor-type protein-tyrosine phosphatase kappa by ultraviolet irradiation activates epidermal growth factor receptor in human keratinocytes. J Biol Chem 2006; 281: 27389-27397.

[98] Kim DJ, Tremblay ML, Digiovanni J. Protein tyrosine phosphatases, TC-PTP, SHP1, and SHP2, cooperate in rapid dephosphorylation of Stat3 in keratinocytes following UVB irradiation. PLoS One 2010; 5: e10290.

[99] Lee H, Morales LD, Slaga TJ, Kim DJ. Activation of T-cell protein-tyrosine phosphatase suppresses keratinocyte survival and proliferation following UVB irradiation. The Journal of Biological Chemistry 2015; 290: 13-24.

[100] Cool DE, Tonks NK, Charbonneau H, Walsh KA, Fischer EH, Krebs EG. cDNA isolated from a human T-cell library encodes a member of the protein-tyrosine-phosphatase family. Proc Natl Acad Sci U S A 1989; 86: 5257-5261.

[101] Mosinger B, Jr., Tillmann U, Westphal H, Tremblay ML. Cloning and characterization of a mouse cDNA encoding a cytoplasmic protein-tyrosine-phosphatase. Proc Natl Acad Sci U S A 1992; 89: 499-503.

[102] Tillmann U, Wagner J, Boerboom D, Westphal H, Tremblay ML. Nuclear localization and cell cycle regulation of a murine protein tyrosine phosphatase. Mol Cell Biol 1994; 14: 3030-3040.

[103] Kamatkar S, Radha V, Nambirajan S, Reddy RS, Swarup G. Two splice variants of a tyrosine phosphatase differ in substrate specificity, DNA binding, and subcellular location. J Biol Chem 1996; 271: 26755-26761.

[104] Bourdeau A, Dube N, Tremblay ML. Cytoplasmic protein tyrosine phosphatases, regulation and function: the roles of PTP1B and TC-PTP. Curr Opin Cell Biol 2005; 17: 203-209.

[105] Lee H, Kim M, Baek M, Morales LD, Jang IS, Slaga TJ *et al*. Targeted disruption of TC-PTP in the proliferative compartment augments STAT3 and AKT signaling and skin tumor development. Sci Rep 2017; 7: 45077.

[106] Baek M, Kim M, Lim JS, Morales LD, Hernandez J, Mummidi S *et al*. Epidermal-specific deletion of TC-PTP promotes UVB-induced epidermal cell survival through the regulation of Flk-1/JNK signaling. Cell Death Dis 2018; 9: 730.

[107] Kim M, Morales LD, Baek M, Slaga TJ, DiGiovanni J, Kim DJ.

UVB-induced nuclear translocation of TC-PTP by AKT/14-3-3sigma axis inhibits keratinocyte survival and proliferation. Oncotarget 2017; 8: 90674-90692.

[108] Kim M, Morales LD, Lee CJ, Olivarez SA, Kim WJ, Hernandez J *et al*. Overexpression of TC-PTP in murine epidermis attenuates skin tumor formation. Oncogene 2020; 39: 4241-4256.

Chapter 8

The Interplay of Key Phospholipid Biosynthetic Enzymes and the Yeast V-ATPase Pump and their Role in Programmed Cell Death

Goldie Libby Sherr and Chang-Hui Shen

Abstract

Exposure of the yeast *Saccharomyces cerevisiae* to environmental stress can influence cell growth, physiology and differentiation, and thus result in a cell's adaptive response. During the course of an adaptive response, the yeast vacuoles play an important role in protecting cells from stress. Vacuoles are dynamic organelles that are similar to lysosomes in mammalian cells. The defect of a lysosome's function may cause various genetic and neurodegenerative diseases. The multi-subunit V-ATPase is the main regulator for vacuolar function and its activity plays a significant role in maintaining pH homeostasis. The V-ATPase is an ATP-driven proton pump which is required for vacuolar acidification. It has also been demonstrated that phospholipid biosynthetic genes might influence vacuolar morphology and function. However, the mechanistic link between phospholipid biosynthetic genes and vacuolar function has not been established. Recent studies have demonstrated that there is a regulatory role of Pah1p, a phospholipid biosynthetic gene, in V-ATPase disassembly and activity. Therefore, in this chapter we will use *Saccharomyces cerevisiae* as a model to discuss how Pah1p affects V-ATPase disassembly and activity and how Pah1p negatively affect vacuolar function. Furthermore, we propose a hypothesis to describe how Pah1p influences vacuolar function and programmed cell death through the regulation of V-ATPase.

Keywords: Phospholipid biosynthesis, Pah1p, programmed cell death, vacuolar activity, V-ATPase, pH homeostasis, Hxk2p

1. Introduction

The yeast vacuole is a crucial and dynamic organelle necessary for the survival of the cell. Adverse environmental factors, such as osmotic stress, toxic metal exposure, and alkaline pH conditions can wreak havoc on normal cellular processes and homeostasis. The vacuole helps play a critical role in protecting the cell via the induction of adaptive stress responses that are upregulated to shield the cell against many of these adverse environmental conditions. Of particular importance are the vacuolar proton translocating ATPase (V-ATPase) pumps. These energy requiring proton pumps are found in the membrane of the vacuole and can help defend the cell against many damaging environmental elements. These multi-subunit ATPase

IntechOpen

complexes span the membrane and actively transport hydrogen ions into the lumen of the organelle. While V-ATPases have been highly characterized in vacuoles they are also present in a number of other organelles and cellular structures including lysosomes, Golgi complexes and the endosomes of eukaryotic cells [1–3]. Furthermore, in mammalian cells, these V-ATPase pumps have been found to be recruited to the cellular membranes of specialized cell types for the purpose of transporting protons across the membrane [1, 2].

Much of the research regarding V-ATPases has been conducted in the model system *Saccharomyces cerevisiae*. The main function of the V-ATPase pump is to acidify the organelle and maintain the internal acidic pH which is required for normal vacuolar function. Thus, the pump hydrolyzes ATP and drives the transport of protons across the membrane from the cytosol and into the lumen of the organelle [4]. The yeast vacuole is, in fact, one of the most highly acidic organelles documented in a cell, along with lysosomes in mammalian cells [4]. Vacuolar pH ranges in acidity from pH 5.0 to 6.5, depending on the specific environmental conditions [4–6]. In lysosomes, the internal pH is even lower, ranging from pH 4.5–5.0 [7]. Given the widespread localization of these pumps in various organelle membranes, V-ATPases are involved in a number of vital roles in cellular homeostasis including protein sorting and secretion, vesicular trafficking and zymogen activation [8]. Additionally, they have also been shown to be involved in endocytic and autophagic processes [1, 2, 9]. Since they play a vital position in maintaining cellular homeostasis, much research has been conducted to better understand just how important these pumps are. This review aims to specifically look at the significant role that V-ATPases play in maintaining pH homeostasis in yeast cells and how they are impacted by Pah1p, a key phosphatidate (PA) phosphatase in the lipid biosynthetic pathway that has been linked to apoptotic mechanisms via the regulation of this pump.

2. V-ATPase pump function

V-ATPase pumps have been shown to be crucial for proper cell function and survival. In fact, research experiments involving the deletion of genes that cause the ubiquitous loss of V-ATPase subunits show that it is lethal in almost all organisms, including *Drosophila* fruit flies and mice [10, 11]. Unlike most organisms however, yeast cells are still viable when V-ATPase subunit encoding genes are deleted. As a result, *Saccharomyces cerevisiae* has been an ideal model organism for studying V-ATPase function [12].

Deletion mutation studies in *S. cerevisiae* have shown that the vacuolar acidification process is impaired when genes that encode V-ATPase subunit are removed. Furthermore, growth phenotypes have indicated that cells with V-ATPase subunit deletions are sensitive to alkaline extracellular pH. While these mutants can grow in an extracellular environment of pH 5, they are unable to grow at pH 7.5. They are also sensitive to high levels of calcium in their growth medium, as well as when they are in the presence of heavy metals and oxidants. They also cannot grow on nonfermentable carbon sources [4, 12].

While V-ATPases are necessary for the vacuoles to maintain their internal acidic pH, they actually play a much more widespread role in the homeostasis of the cell by regulating cytoplasmic pH. Research has shown that cells with defective V-ATPase pumps are not only unable to maintain the acidity in their vacuoles but cannot uphold the necessary pH of the cell's cytoplasm [5, 13]. This is due to the fact that other pumps are dependent on V-ATPase activity which utilizes ATP to transport hydrogen ions into the vacuole. For instance, the Ca^{+2}/H^{+} antiporter requires a functional V-ATPase to transport calcium into the vacuole. If the V-ATPase is defective, calcium will enter the vacuole at a much slower rate and thus these mutants cannot grow well in presence of excess calcium as mentioned above [14–16].

Additionally, the vacuole plays a large role in detoxifying the cell of heavy metals via antiporters in the vacuole membrane. These pumps are also dependent on functional ATPases and rely on their activity to function properly. Cells with non-functional V-ATPases are unable to thrive in the presence of heavy metals, such as cadmium which causes oxidative stress in the cell [17, 18]. Furthermore, V-ATPases play an important part in promoting hydrolytic enzymes that degrade and recycle biomolecules [9, 19]. When a cell undergoes nitrogen starvation, nearly 4/5ths of the cell's protein degradation process takes place in the vacuole. Additionally, the vacuoles recycle organelles as well, which also helps to supply the cells with an abundance of amino acids and other biomolecule building blocks that can help the cell survive during times of stress [20–22]. V-ATPases therefore play a critical role in protecting the cell when exposed to various adverse environmental conditions and stress.

3. Structural composition of the vacuolar V-ATPase pump

Given the importance of V-ATPase pumps in maintaining intravacuolar acidity, the structural components of the pump have been well characterized in *Saccharomyces cerevisiae*. V-ATPase pumps are comprised of two distinctive domains that include the V1 domain and the V0 domain. The V1 domain is the catalytic component in the pump that is linked to the ATP binding sites and is located on the cytosolic part of the membrane. The V0 domain is the proton translocating component and consists of integral membrane proteins as well as peripheral ones. Both the V1 and V0 domains are linked together by a stalk-like structure within the V1 domain [13, 23]. These domains each consist of multiple subunits that come together to form the overall pump. V1 contains eight subunits, labeled A through H (A, B, C, D, E, F, G, and H) while the V0 domain contains 6 subunits labeled a, c, c', c", d and e (**Figure 1**).

The 14 subunits of the V-ATPase pump are each encoded by their own gene, except for subunit a of the V0 domain which has two isoforms. 13 of these subunits are encoded by vacuolar membrane ATPase (VMA) genes [4]. These genes include *VMA1, VMA2, VMA3, VMA4, VMA5, VMA6, VMA7, VMA8, VMA9, VMA10, VMA11, VMA13, VMA16* (**Table 1**). The final subunit in the vacuolar V-ATPase is encoded by a gene named *VPH1* and encodes for the largest subunit in the pump, subunit a [15]. However, this subunit has two isoforms, with the second isoform being encoded by *STV1*. The Stv1p isoform is not located in the vacuolar membrane but is instead found in the V-ATPases located in the Golgi complex and endosome membranes. Research has shown that vacuolar V-ATPases that contain the Vph1p subunit are better at coupling the hydrolysis of ATP to the transport of hydrogen ions across the membrane. They also have increased assembly levels of the V-ATPase pump, and are more responsive to extracellular glucose levels compared to V-ATPases that contain Stv1p [4, 24]. For the purposes of this review, we will be focusing on vacuolar V-ATPases and will be looking at the Vph1p subunit.

Thus, eight of the V-ATPase subunits in the vacuole are found in the V1 domain and are encoded by *VMA1, VMA2, VMA4, VMA5, VMA7, VMA8, VMA10, VMA13*. The remaining six subunits are encoded by *VPH1, VMA3, VMA6, VMA9, VMA11*, and *VMA16* and are found in the V0 domain. A table outlining the various genes of the V-ATPase pump and the subunits encoded by them has been included (**Table 1**).

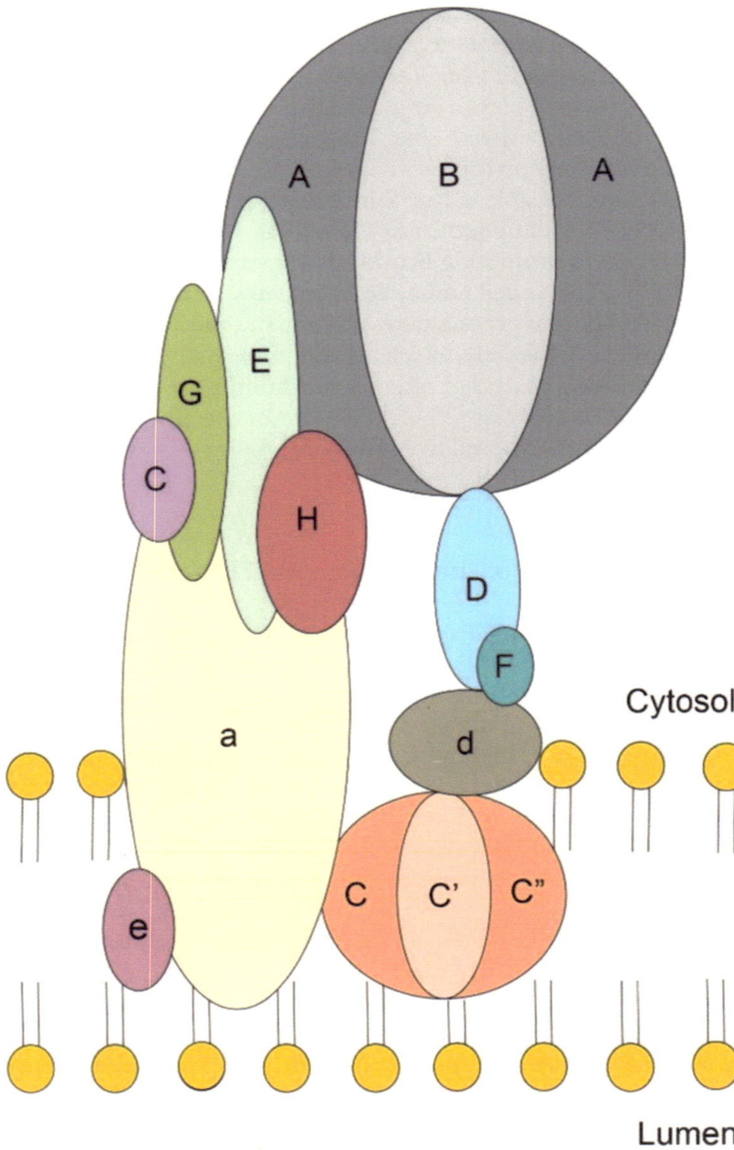

Figure 1.
Model of the Saccharomyces cerevisiae *vacuolar V-ATPase. The V1 domain contains subunits A, B, C, D, E, F, G, and H. Subunits A and B each have three copies (not all depicted) and alternate in an A, B, A, B, A, B fashion in the pump. The Vo domain consists of subunits a, c, c', c", d, and e.*

Domain	Gene	Subunit
V_1	VMA_1	A
V_1	VMA_2	B
V_1	VMA_4	E
V_1	VMA_5	C
V_1	VMA_7	F
V_1	VMA_8	D
V_1	VMA_{10}	G

Domain	Gene	Subunit
V_1	*VMA*$_{13}$	H
V_0	*VPH*$_1$	a
V_0	*VMA*$_3$	c
V_0	*VMA*$_6$	d
V_0	*VMA*$_9$	e
V_0	*VMA*$_{11}$	c′
V_0	*VMA*$_{16}$	c″

Table 1.
The genes of the Saccharomyces cerevisiae *vacuolar V-ATPase and their corresponding subunits.*

4. V-ATPase subunit functions

The overall molecular mechanism of V-ATPase pump activity involves the hydrolysis of ATP by the V1 domain. This step provides the necessary energy needed to pump protons across the membrane and create a pH gradient that is needed for secondary transport systems [3, 25]. However, the specific functions of each of the subunits of the V1 and V0 domains are varied. The first two subunits of the V1 domain, subunit A and subunit B, of which there are three copies of each, are encoded by *VMA1* and *VMA2* respectively. They are the ATP binding subunits that hydrolyze ATP. Additionally, both aid in the regulation of proton transport and in the disassembly of the V1 and V0 domains [26, 27]. Subunit C, which is encoded by the *VMA5* gene, is needed to help activate the pump by regulating the assembly of the V1 domain to the V0 domain. Furthermore, it plays a key role in the disassembly of the pump since it dissociates from the enzyme during glucose starvation and separates the two domains, rendering the pump inactive [28, 29]. Subunit D is encoded by *VMA8* and is also crucial for the assembly of the two domains which is needed for pump activity and proton transport [30]. Subunit E, encoded by the *VMA4* gene, helps to form the peripheral structural stalk of the pump which is needed for proper assembly and function of the pump, while subunit F which is encoded by *VMA7* is a rotor subunit and needed for proper pump assembly [31–34]. Subunit G is encoded by *VMA10* and needed for proper stalk formation, while subunit H, which is encoded by *VMA13*, is required to inhibit ATP hydrolysis when V1 and V0 domains are dissociated and does so by interacting with subunit F [32, 35].

The subunits of the V0 domain are the remaining components of the V-ATPase pump that are primarily integral proteins, with the exception of subunit d which is a peripheral protein. Subunit a is encoded by the *VPH1* gene which are specific for V-ATPases localized in the vacuolar membrane. It is needed for the appropriate assembly of the pump as well as for the transport of hydrogen ions [24, 36]. Subunits c, c' and c", which are encoded by *VMA3, VMA11,* and *VMA13* respectively, are suggested to be needed for proton translocation. All three subunits are crucial for V-ATPase function since a mutation in any of them will impair pump activity [37, 38]. Subunit d, which is encoded by *VMA6*, is the only peripheral V0 subunit and plays a key role in coupling ATP hydrolysis with proton transport [39]. Lastly, it is also important to note that the e subunit in the V0 domain, which is encoded by *VMA9*, was discovered much later than the rest of the subunits and as such has not been as well characterized [12]. Furthermore, *in vitro* studies have suggested that Vma9p is not needed for V-ATPase proton pumping activity since removal of Vma9p does not impact proton transport [40].

Importantly, the deletion of any one specific subunit in the V1 domain does not impact the stability of the remaining V1 subunits in the complex. However, it will impair the association and assembly of the entire V1 domain with the V0 domain, thus rendering the pump nonfunctional. Additionally, the deletion of any specific V1 subunit will not impact the stability of the V0 domain [13, 41]. Likewise, the loss of any one specific V0 subunit will not impact the stability of the remaining V0 subunits nor will it impact the assembly of the independent V1 domain. It will however impact the assembly of the V0 domain which is unable to produce a functional V-ATPase [13, 41]. It is important to note that if any of the thirteen genes that encode the V-ATPase pump are implicated (not including *VMA9*), the pump will not be able to work properly. Thus, if there is a mutation in any single subunit from either domain, the cell will be unable to grow in media whose pH is neutral or basic [42].

5. V-ATPase mechanism of disassembly and assembly

The disassembly of the V-ATPase is a crucial step required for regulating the activity of the pump. This mechanism involves the separation of the V1 and V0 domains and will ultimately inhibit the V-ATPase pump's function *in vivo*. When undergoing disassembly, the C subunit located in the V1 domain, which acts as a bridge between the V1 and V0 domains, will depart from the complex and cause the separation of the two domains [43, 44]. Subsequently, a conformational change in the H subunit inhibits ATP hydrolysis from occurring in the newly separated V1 complex. This occurs so that energy will not be unnecessarily used without the concurrent transportation of protons across the membrane [45–48]. Furthermore, once the domains have disassembled, the passive transport of hydrogen ions across the V0 complex is also prevented [49].

This mechanism is completely reversible however once the C subunit is brought back and re-bridges the V1 and V0 domains [43]. In *Saccharomyces cerevisiae*, re-assembling the V-ATPase pump utilizes a chaperone protein that is specific for V-ATPase pumps [50]. Specifically, this process requires RAVE, or the regulator of ATPases of Vacuoles and Endosome [3]. This is a chaperone complex that includes an adaptor protein, Skp1p and the functional subunits Rav1p and Rav2p [50, 51]. This RAVE chaperone complex helps to stabilize the V1 domain and facilitates in the process of reintroducing the C subunit to bridge V1 to V0 [52, 53]. Once completed, the V-ATPase pump will be fully functional once again.

6. Importance of V-ATPase activity in mammalian cells

The study of vacuolar function in *Saccharomyces cerevisiae*, and particularly the activity of the V-ATPase pump is important since yeast vacuoles are strikingly similar to the lysosomes in mammalian cells [52. 53]. The mammalian lysosome has an internal acidic pH that ranges from 4.2–5.3 and is primarily maintained by V-ATPase pumps that actively transport ions across the membrane to acidify the lysosome [54]. Similar to the acidic pH needed in the vacuole, the maintenance of lysosomal pH is required for the important hydrolytic activities and signaling roles of lysosomal homeostasis. Defects in V-ATPase machineries and functions have been associated with a number of neurodegenerative diseases, especially disorders associated with older age including some forms of Parkinson's disease and Alzheimer's disease [55–57]. In fact, the brain is one of the main organs that is the most significantly impacted by genetic diseases that interrupt normal lysosomal function in cells. This underlies the importance of normal lysosome function in the central nervous system [54, 58]. Additionally, impairments to proper acidification in lysosomes have been involved with cell aging and longevity. Studies conducted in yeast models have shown that longer lifespans occurred when V-ATPase components were overexpressed [59, 60]. Additionally, there have been over fifty genetic disorders that have been linked to mutated genes that encode specific proteins of the lysosome, many of which are related to the acidification function [61–63].

V-ATPases have been found in the cellular membranes of certain specialized cells in mammals. Implications in these V-ATPases cause an array of genetic disorders. For example, V-ATPases that are found in intercalated cells of the distal tubule and the collecting ducts of the kidney play a crucial role in maintaining acid–base homoeostasis. In humans, defects in the specific genes that encode V-ATPase subunits, such as a mutation in the renal isoform of subunit B or in subunit a, leads to the inherited disease of renal tubule acidosis [64, 65]. Furthermore, V-ATPase pumps have also been found to be located in the cellular membrane of osteoclast cells and are required for the process of bone resorption. As a result, genetic mutations in V-ATPase encoding genes, such as in subunit a, has been associated with osteopetrosis, which causes skeletal abnormalities caused by lack of bone degradation [64, 65]. Furthermore, V-ATPase activity has been linked to nongenetic diseases, such as cancer. The key role that these pumps play in tumor and cancer cell lines will be discussed further in a later section. Given the importance of V-ATPases with regard to diseases, research in yeast vacuoles has been crucial in helping better understand the role that these pumps play in cellular homeostasis.

7. Regulation of the lipid biosynthetic pathway

The relationship between V-ATPases and the lipid biosynthetic pathway is only recently becoming better understood. The lipid biosynthetic pathway is a highly controlled pathway that synthesizes the crucial membrane phospholipids, in addition to other various fats and lipids that are necessary for cell survival [66–68]. The transcription of genes in this pathway is tightly governed by a number of key regulators. One of these primary regulators is inositol, a crucial phospholipid precursor that plays a key role in regulating phospholipid metabolism based on its availability [69–72]. When present in the growth medium, inositol is not required to be synthesized by the cell and thus the genes involved in its production are turned off. If inositol is absent in the growth medium, then the cell will need to produce the crucial phospholipid precursor. Multiple genes in the phospholipid biosynthetic pathway have a UAS_{INO}, or an inositol responsive cis-acting element, that is located in their promoter regions [73, 74]. This region is the binding site for the Ino2p/Ino4p activator which activates the transcription of genes needed to produce inositol. Conversely, when inositol is richly present in the media, the Opi1p repressor will bind to the Ino2p/Ino4p activator and thus repress transcription from occurring [74, 75].

Another crucial regulator of the lipid biosynthetic pathway is the phosphatidate (PA) phosphatase enzyme, Pah1p. Pah1p, which is encoded by the *PAH1* gene, catalyzes the crucial reaction that dephosphorylates PA and leads to the production of diacylglycerol (DAG) and a phosphate group in the lipid biosynthetic pathway [76]. Both PA and DAG are two central players in this pathway and their levels are highly regulated, which in turn regulates the synthesis of triacylglycerides and membrane phospholipids [77]. Depending on the pathway taken, DAG can either be the precursor of the phospholipids phosphotidylcholine and phosphatidylethanolamine or can be converted into triacylglycerol (TAG). PA on the other hand can generate all four major phospholipids that are found in the membrane of *Saccharomyces cerevisiae*. If it is not dephosphorylated and converted into DAG, then PA can be used to manufacture phosphatidylinositol and phosphotidylserine and can also be used to produce phosphotidylcholine and phosphatidylethanolamine [77].

Since PA and DAG play such critical roles in the lipid biosynthetic pathway, they are both highly regulated by the key PA phosphatase enzyme, Pah1p. While there are other PA phosphatase enzymes, Pah1p is considered to participate more directly in the synthesis of phospholipids and plays a pivotal regulatory role in the pathway [78, 79] and is essential for the *de novo* synthesis of membrane phospholipids and TAG [80, 81]. Pah1p is mainly located in the cytosol where it can be easily translocated onto the endoplasmic reticulum to catalyze the conversion of PA into DAG [82]. However, Pah1p's regulatory role in the lipid biosynthetic pathway extends even further due to the fact that it exerts transcriptional regulation over other genes involved in the manufacturing of phospholipids. As explained earlier, numerous genes in the lipid biosynthetic pathway contain a UAS_{INO} element in their promoter regions which acts as a binding site for the transcriptional activator, Ino2p/Ino4p [83, 84]. The levels of PA actually help control the transcription of UAS_{INO} containing genes, since Opi1p, which is a negative regulator of transcription, is physically connected to PA on the endoplasmic reticulum and nuclear membrane [85]. Thus, when there are greater levels of PA, the Opi1p repressor will continue to remain tethered to PA on the ER/nuclear membrane which will prevent it from being able to cross into the nucleus and prevent the transcription of UAS_{INO} containing genes [86]. Conversely, when PA levels are decreased, Opi1p is free to cross the nuclear membrane and bind to the Ino2p/Ino4p activator complex on the UAS_{INO} and block transcription. Since Pah1p regulates PA levels by catalyzing the reaction to turn PA into DAG, it indirectly impacts the transcription of other lipid biosynthetic genes. Therefore, higher concentrations of Pah1p will lead to less PA which represses transcription of genes with a UAS_{INO}, while lower concentrations of Pah1p leads to increased amounts of PA and activates transcription of genes with a UAS_{INO} [86]. However, additional research has uncovered that Pah1p actually plays a more direct role in the regulation of lipid biosynthetic genes as well. Studies have found that Pah1p is located in the nucleus as well so that it can directly act as a repressor for UAS_{INO} genes in the pathway [83]. In fact, it can directly bind to the promoter of UAS_{INO} containing genes and physically block gene expression. It is therefore not surprising that studies looking into the impact of deleting the *PAH1* gene show that there is an upregulation of gene expression of UAS_{INO} containing genes [87].

In addition to its regulatory role, deletion experiments have shown just how pivotal Pah1p is to overall cell homeostasis. Mutants that lack the *PAH1* gene have much higher levels of PA present in the cell. Additionally, these cells contain much lower levels of TAG due to the loss of the Pah1p phosphatase activity and the conversion of PA into DAG, which is the precursor of TAG. Furthermore, there is an abnormal expansion of the nuclear and endoplasmic reticulum membrane as well as increased levels of phospholipids, sterol esters and fatty acids in cell lines lacking *PAH1* [88, 89]. This is because, the gene expression of UAS_{INO} genes in the lipid biosynthetic pathway is upregulated. This is due to the lack of Pah1p which causes a derepression of genes that are typically repressed in the presence of Pah1p in the CDP-DAG pathway and Kennedy pathways, both of which lead to the synthesis of membrane phospholipids [87]. These mutant strains also experience fatty acid toxicity due to the higher levels of lipids present [89]. *PAH1* has also been shown to be needed for lipid droplet formation, which is dependent on the presence of DAG. Thus, in the absence of *PAH1* and the resulting lower concentrations of DAG, there is a decrease in the concentration of lipid droplets in these cell [90]. Furthermore, cells that are missing *PAH1* are not able to grow in the presence of non-fermentable carbon and are also temperature sensitive [91]. Most notably for this review article, *PAH1* has been shown to be important for vacuole morphology and function. Given the importance of V-ATPase activity in overall cell homeostasis, the link between Pah1p and vacuole function is important to understand and will now be explored in the upcoming sections.

8. The role of *PAH1* in vacuolar morphology and function

Interestingly, studies have revealed the importance of the *PAH1* gene in vacuolar homeostasis and morphology. Research has shown that cells lacking *PAH1* have morphologically defective vacuoles that remain interminably fragmented [92, 93]. While vacuoles typically undergo fragmentation in *Saccharomyces cerevisiae* during the budding process so that the new daughter cell can obtain and inherit these organelles, the vacuoles do not remain fragmented indefinitely [94]. This normal mechanism of fragmentation begins when the budding process is initiated and the vacuole in the mother cell fragments into a series of mini vacuoles. This is followed by the development of an elongated-vesicular structure by the mini vacuoles, which guides the structures to the newly budding yeast cell [94]. Once these mini fragmented vacuoles are inherited and the newly budding cell grows, the collection of vacuoles fuse together to recreate a single large vacuole in both the mother cell and daughter cell. The two cells are separated once septa are formed and this cycle can be repeated as more daughter cells are made [95]. However, cells that lack *PAH1* are unable to keep their vacuoles unfragmented and whole [92, 93]. Research studies have found that this is due to the fusion machinery having been implicated when *PAH1* is deleted and thus prevents the vacuoles from fusing back together. Without the enzymatic phosphatase activity that exists when *PAH1* is present, the SNARES are unable to bind to Sec18p, which is the protein required to prime the SNARE complex for the fusion mechanism. Furthermore, the deletion of *PAH1* also causes a number of other key fusion machinery components to be absent. This includes Vps39p, which is a component of the homotypic fusion and vacuole protein sorting (HOPS) tethering complex. Additionally, phosphatidylinositol 3-phosphate, a lipid that is needed for SNARE function and fusion, is also absent [92]. Moreover, Pah1p is also needed to recruit, Vps34p, which is the phosphatidylinositol 3-kinase needed for vacuolar fusion [92]. As such, vacuoles in cells that lack *PAH1* remain fragmented without ever fusing into a single vacuole.

Since irregularities in vacuole morphology have been linked to implications in V-ATPase pump activity, there is reason to suspect a relationship between Pah1p and V-ATPase activity given the important role that Pah1p plays in vacuolar morphology [93, 94]. On the other hand, reports have found that vacuoles with V-ATPase pumps that have a decrease in their level of acidifying the vacuole can actually lead to increased fusion of the vacuoles. One particular study found that vacuoles that experience deacidification often have an increased level of fusion whereas mutant vacuoles with internal pHs that have maintained its acidity may have fusion impeded [96]. Thus, the uncertainty as to whether the role of Pah1p and its impact on the morphological structure of vacuoles was related to proper V-ATPase pump function remained.

Given the fact that Pah1p regulates genes in the lipid biosynthetic pathway, as well as helps play a critical role in the maintenance of proper vacuole morphology and function, additional studies were aimed to look at whether Pah1p plays a role in regulating the genes that encode for the V-ATPase pump and its impact on pump activity. Growth experiments have shown that cells lines lacking *PAH1* actually grew better in neutral environments compared to wildtype. Since neutral environments can impact cytoplasmic and vacuolar pH, a properly functioning V-ATPase pump is required to ensure that pH homeostasis is maintained in the vacuole and the entire cell, indicating that *pah1Δ* cells did not negatively impact V-ATPase activity even though the morphological structure is affected. This was proven with subsequent experimentation that measured vacuolar pH and actually showed that *pah1Δ* cells were even better than wildtype cells at acidifying their vacuoles, with an average internal vacuolar pH of 5.89 in their vacuoles compared to pH 6.0 in wildtype. Therefore, while Pah1p does cause morphological disturbances to the vacuole it does not adversely impact V-ATPase pump activity. This however is not a contradiction since cell lines that have mutations that cause abnormalities in vacuole morphology and fragmentation can either have fully functioning V-ATPase pumps or conversely pumps that have defects in their assembly [97, 98].

Interestingly, other research studies have shown that V-ATPase function can adversely impact the fusion of vacuoles *in vivo* [96]. Therefore, it is likely that the increased acidity in the *pah1Δ* cells contribute to the fragmentation of the vacuole seen in these cell lines. This led to the discovery that perhaps Pah1p plays a regulatory role over V-ATPase genes as well, since pump activity was upregulated in the *pah1Δ* strain. RNA analysis experimentation showed that 11 of the 13 vacuolar membrane ATPase genes were upregulated in the V-ATPase pump, including *VMA3, VMA6 and VMA16* which are all involved in the transport of hydrogen ions into the vacuole [93]. This indicates a potential role of Pah1p acting as a repressor for these genes since they are seemingly negatively regulated in the presence of Pah1p. Many of the 11 genes that were impacted by the deletion of PAH1p contain a UAS_{INO}, which as mentioned earlier are types of genes that Pah1p negatively regulates in the lipid biosynthetic pathway. The *VMA* genes that possess a UAS_{INO} in their promoters include *VMA1, VMA5, VMA8, VMA13* and *VMA16* [93]. There is therefore a molecular relationship between *PAH1* and the genes that encode V-ATPases since it appears that Pah1p can directly negatively impact these genes by binding to their promoters.

9. V-ATPases and glucose metabolism

One important process that V-ATPase pumps have been found to be associated with is glucose metabolism. In fact, research has shown that V-ATPase pump activity is actually regulated by glucose. In both yeast and mammalian cells, a key inducer of the V-ATPase pump disassembly is glucose depletion [20, 43, 99]. When glucose is scarce, the V1 and V0 domains will disassemble and pump activity is unable to occur. Conversely, an increase in glucose levels, which triggers the activation of glycolysis, will lead to the reassembly of the V-ATPase pump and lead to an increase in pump activity. This has been found to occur in both yeast cells [8, 43] and mammalian cells [99, 100]. This process is extremely significant since during times of glucose depletion the cell aims to preserve energy. By disassembling the V-ATPase pump, it prevents unnecessary usage of ATP. Furthermore, in times of glucose abundance, the reassembling of the pump allows for a functioning V-ATPase that can lower the additional acidification of the cytosol that occurs during an uptick in glycolysis [1, 2, 25]. Research has shown that this cycle of the V-ATPase pump being disassembled and reassembled is proportional to the concentration of glucose available in the cell. This is important because it indicates that V-ATPase pumps are highly attuned to glucose metabolism and energy levels [8, 43, 101, 102].

In yeast, the highly characterized glucose sensing signaling mechanism which regulates the assembly of the V-ATPase pump is the Ras/cAMP/Protein Kinase A

(PKA) pathway [3, 103]. The GTP coupled protein, Ras, is inhibited by two GTPase activating proteins named Ira1p and Ira2p. When glucose is present, Ira1p and Ira2p are inhibited and Ras can stimulate the production of cAMP via adenylate cyclase. Once cAMP levels are high enough, the PKA regulatory subunit will dissociate and be free to trigger PKA's kinase activity. Additionally, studies have indicated that the assembly of the V-ATPase takes place as a result of acidification of the cytosol due to high levels of glycolysis and that this leads to changes in PKA [104, 105]. The presence of glucose, after a period of depletion, will activate PKA and thereby stimulate the assembly of the V-ATPase pump. This mechanism seemingly creates a positive feedback loop, since the heightened assembly of V-ATPase pumps aids in the maintenance of the pH in the cytosol and can stimulate PKA signaling. This in turn will cause an upregulation in glycolysis which further boosts the assembly of V-ATPase pumps and helps facilitate the switch from respiratory to fermentative growth [3, 25].

10. V-ATPase and cancer

As mentioned earlier, V-ATPase pump activity and function has been associated with various diseases. However, one such non genetic disease that these pumps have been linked to is cancer [106]. Normal V-ATPase pump functioning is crucial for various signaling pathways in the cell. In fact, many of the pathways that lose their homeostatic control during cancer require V-ATPase pumps for proper functioning, such as the Notch signaling pathway and the Wnt/β-catenin signaling pathway [106–108]. Moreover, proper pump activity is required for the activation of the mechanistic target of rapamycin complex, or mTORC1, which is a common pathway implicated in cancer [109–113]. However, studies have shown more direct roles of V-ATPase pumps in cancer. For example, when V-ATPase pumps are inhibited in human tumor cells, results showed programmed cell death is induced [114–119]. These findings indicate that cancer cells rely on V-ATPase activity much more than noncancerous cells do in order to remain viable [115, 116, 119]. This is because cancerous cells have been shown to generate increased levels of acidity [120]. Since V-ATPase pumps play the critical role of removing acids out of the cell and help to preserve a neutral pH in the cytosol, cancer cells heavily rely upon V-ATPases to promote the alkalinization of their cytosol and to pump protons into the extracellular space [1, 2, 121]. It has been hypothesized that expression of these pumps may upregulate or may be specifically localized to the cellular membrane by expressing certain isoforms that target the V-ATPases to the plasma membrane [64, 65, 122, 123]. In this way, tumors are thus able to dodge programmed cell death and can proliferate [106]. In fact, V-ATPase pumps have been found in the plasma membrane of numerous invasive cancer cell lines while not in non invasive ones [124]. For example, an invasive line of breast tumor cells revealed higher levels of the V-ATPase pump on the cell surface compared to noninvasive breast cancer cells [124]. As explained above, these pumps aid in the maintenance of cytosolic pH. The impact of V-ATPases has also been observed in various other invasive cancerous cell lines, including liver cells, esophageal cells, ovarian cells, lung cells, prostate cells and pancreas cells amongst others [106, 110, 111, 125–129]. Research has shown that inhibiting V-ATPase pumps disrupts the pH balance and causes a more acidic pH in the cell which leads to higher levels of apoptosis [130].

Furthermore, V-ATPase pumps are not only important for regulating the pH of the cytosol, but for regulating the pH of other organelles as well [106]. Implications of protein production and shortages in nutrients can lead to ER stress in cancer cells. As a result, lysosomes are activated by the Bax inhibtor-1 in response to the

ER stress in order to raise protein turnover [131]. Studies have shown that blocking ATPase pump activity barred this from occurring and ultimately led to cell death [131]. Moreover, V-ATPases in the cell membrane have also been proposed to play a crucial role in the migration and invasion of cancer cells. Studies have shown that when V-ATPase pumps were inhibited in invasive breast cancer cells, the migration and invasiveness of these cells was diminished [1, 2, 124]. This was further proven in other cancer lines as well, such as cancerous pancreas cells [125]. Thus, the relationship between V-ATPases and cancer, and subsequently its link to apoptosis, has been an active area of research. However, the role of V-ATPases in yeast and its connection to apoptosis has not been as well characterized compared to that of human cell lines. This review will now look at a possible mechanism by which V-ATPase activity can be linked to apoptosis via its regulation by key players of the lipid biosynthetic pathway.

11. V-ATPase genes and the lipid biosynthetic pathway

As explained earlier, inositol is one of the key phospholipid precursors whose presence is essential in regulating phospholipid metabolism [69–72]. When present in the growth medium, inositol production is repressed. If inositol is absent in the growth medium, then the genes involved in its production are upregulated. Research has shown that cells that lack one or more of the vacuolar membrane ATPase genes exhibited defects in growth when cultured in media without inositol [132, 133]. However, these defects were able to be reversed if *OPI1*, the gene that encodes the Opi1p repressor, was deleted [133]. Additionally, cells that lack one or more V-ATPase genes have higher buildups of oxidant molecules which may indicate their relationship to protecting the cell and their role in preventing cell death.

Recent studies have shown that one of the key vacuolar membrane ATPase genes, *VMA3*, plays a significant role in regulating the synthesis of phospholipids. Growth experiments that were performed with cell lines lacking the *VMA3* gene exhibited a growth defect that had cells growing much slower when cultured without inositol in their media compared to wildtype cells [134]. *VMA3*'s role in *de novo* phospholipid production was further clarified when mRNA studies showed *HXK2*, another important gene in the phospholipid biosynthetic pathway required for inositol production, had significantly lower mRNA levels in the cells lacking *VMA3* compared to wildtype. Thus, *VMA3* was revealed to impact the phospholipid biosynthetic pathway by specifically regulating the transcription of the *HXK2* gene [134]. This finding was particularly interesting since other studies have shown that cells lacking *HXK2* exhibit a growth sensitivity to acetic acid [135], which is a growth condition that has long been used to screen for apoptosis. Earlier studies have hypothesized that *HXK2* plays a role in shielding cells from programed cell death since cells that have had the *HXK2* gene deleted had an accrual of activated Ras by the mitochondria [135]. Thus, further experimentation from this study showed that *vma3Δ* cells grew significantly slower in the presence of acetic acid compared to wildtype and were even more sensitive to this apoptotic inducing agent when grown without inositol present. Taken together, these findings indicate that the deletion of *VMA3* leads to decreased transcription of the *HXK2* gene, which ultimately leads to cells being more sensitive to acetic acid. This therefore demonstrates that *VMA3* plays an important regulatory role in apoptosis [134].

HXK2 plays an important role in glucose metabolism and encodes the hexokinase-2 enzyme that catalyzes the reaction which converts glucose into glucose-6-phospate. Thus, *HXK2* is regulated by the presence of glucose. Studies have shown that *HXK2* is regulated by two important transcription factors, Rgt1 and

Med8, that cause the deregulation of *HXK2* when glucose is not present [136]. The findings by Konarzweska *et al.* have now also shown that *HXK2* is also regulated by Vma3p. Thus, there is a clear relationship between V-ATPases and the phospholipid biosynthetic pathway since Vma3p has been shown to upregulate the *HXK2* gene. Furthermore, it has been shown that by regulating the *HXK2* gene, *VMA3* has an important protective role when it comes to acetic acid induced apoptosis. This is therefore an important link between how V-ATPases relate to the lipid biosynthetic pathway.

12. A working model

As first mentioned in a 2017 article by Konarzewska *et al.*, a new model is now being hypothesized as to how Pah1p impacts apoptosis by regulating the V-ATPase *VMA3* gene, which in turn has a regulatory role over *HXK2* (**Figure 2**). This model starts with the availability of glucose that leads to a cytosolic pH, which creates a more alkaline internal environment, and will also activate protein kinase A [25]. This in turn causes the assembly of V-ATPase machinery and the formation of a functional V-ATPase pump. During this time, PA levels remain high while Pah1p's concentration is low. When PA is abundant due to lower Pah1p levels, it will follow the CDP-DAG pathway and create membrane phospholipids. Since V-ATPase assembly is being promoted, the VMA genes that encode these subunits are upregulated. Importantly, one of *VMA* genes yields Vma3p which is a key player in this model. Additionally, the lipid Phosphatidylinositol 3,5-bisphosphate can also help in the activation of the V-ATPase pump and as more V-ATPase pumps assemble and accumulate the alkaline cytosolic pH will be upheld [137]. Protein kinase A signaling can be stimulated by the newly assembled V-ATPase pumps, which leads to a positive feedback loop and increases the numbers of V-ATPase pumps being assembled. Additionally, there will be an upregulation of glycolysis which will lead to an abrupt changeover to fermentative growth. Regardless of whether the Ras/cAMP/PKA pathway is upstream or downstream of the V-ATPase pump assembly pathway [104, 134], Ras will be largely found by the cell membrane and inside the nucleus. Furthermore, the *HXK2* gene will exhibit maximal expression due to the high levels of glucose and the gene product Hxk2p will catalyze the reaction to convert glucose into Glucose-6-phosphate. Given the role of Hxk2p in apoptosis, the cell will have a greater resistance against acetic-acid induced apoptosis.

Conversely, when glucose is not abundantly present, the machinery for the V-ATPase pumps will undergo disassembly due to the less alkaline cytosolic environment. The *PAH1* gene will be upregulated and thus result in increased production of the active Pah1p which can execute its phosphatase activity on PA. This is turn will cause PA to be led along the pathway that converts it to DAG and ultimately TAG, thus minimizing the production of membrane phospholipids from PA. Furthermore, the increased amounts of Pah1p will repress most vacuolar membrane ATPase gene expression which also prevents further V-ATPase assembly and enhances its disassembly. Being that one of the *VMA* genes that Pah1p represses is *VMA3*, there will be less Vma3p available. Vma3p's regulation over *HXK2* leads to a decrease in Hxk2p. During this time, Ras will build up in the internal membranes and mitochondria which can ultimately cause dysfunction of the mitochondria and amplified ROS production [138]. This will therefore make cells more susceptible to acetic acid induced programmed cell death. Thus, Pah1p likely plays a major role in this apoptotic pathway given its regulatory role over Vma3p.

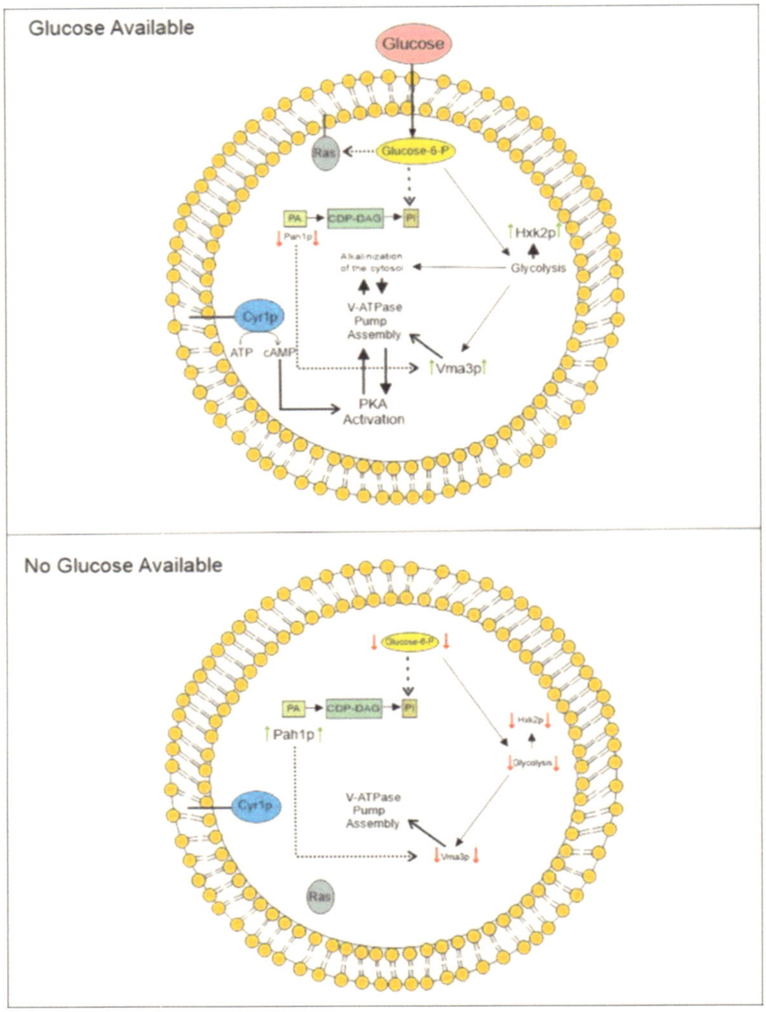

Figure 2.
Proposed model first hypothesized in [134] for Pah1p as an apoptotic regulator via the regulation of Vma3p, which subsequently regulates Hxkp2. When glucose is available, V-ATPase pump assembly occurs due to the activation of PKA and cytosolic alkalinization as a result of glycolysis. Concurrently, Pah1p levels are decreased, which promotes the CDP-DAG pathway and PA being converted to membrane phospholipids. Vma3p, amongst other VMA genes, is upregulated due to lack of Pah1p and the increased levels of pump assembly uphold the cytosolic alkalinization. Furthermore, this leads to increased Hxk2p expression which results in the resistance of acetic acid induced apoptosis. Conversely, if glucose is not available then V-ATPase pumps undergo disassembly. Pah1p levels are high, which catalyzes the reaction of PA conversion to DAG which ultimately increases TAG levels. Furthermore, there is downregulation of VMA3, along with other VMA genes. This leads to decreased pump assembly, low levels of Hxk2p and an increase of Ras in the mitochondria and membranes. Ultimately, acetic acid induced apoptosis is promoted under these conditions [134].

13. Conclusion

Overall, this review shows the connection between the V-ATPase pump and the lipid biosynthetic pathway PA phosphatase regulator, Pah1p. Based on the regulatory role that Pah1p has over the *VMA* genes and specifically the *VMA3* gene, a new model has emerged regarding how it can possibly influence acetic acid induced apoptosis through the regulation of V-ATPase genes. While more research studies will be needed to confirm this model, these studies have indicated how Vma3p can act as potential anti apoptotic factor in *Saccharomyces cerevisiae.* These findings thus highlight the importance Pah1p on vacuolar function and on cell induced apoptosis via Vma3p.

Acknowledgements

We are grateful to our laboratory colleagues for technical assistance and comments on the manuscript. This work was supported by a NATO Grant (NATO SPS G5266), PSC-CUNY awards, and CUNY Institute for Macromolecular Assemblies research support to C.-H. S.

Author details

Goldie Libby Sherr[1] and Chang-Hui Shen[2,3,4,5]*

1 Department of Biological Sciences, Bronx Community College, City University of New York, NY, United States

2 Department of Biology, College of Staten Island, City University of New York, Staten Island, NY, United States

3 Biology Program, The Graduate Center, City University of New York, NY, United States

4 Biochemistry Program, The Graduate Center, City University of New York, NY, United States

5 Institute for Macromolecular Assemblies, City University of New York, NY, United States

*Address all correspondence to: changhui.shen@csi.cuny.edu

References

[1] Cotter, K., Capecci, J., Sennoune, S., Huss, M., Maier, M., Martinez-Zaguilan, R., & Forgac, M. (2015). Activity of plasma membrane V-ATPases is critical for the invasion of MDA-MB231 breast cancer cells. Journal of Biological Chemistry, *290*(6), 3680-3692.

[2] Cotter, K., Stransky, L., and Forgac, M. (2015). Recent insights into the structure, regulation, and function of the V-ATPases. Trends Biochem. Sci. 40, 611-622. doi: 10.1016/j.tibs.2015.08.005

[3] Hayek, S. R., Rane, H. S., & Parra, K. J. (2019). Reciprocal regulation of V-ATPase and glycolytic pathway elements in health and disease. Frontiers in physiology, *10*, 127.

[4] Li, S. C., & Kane, P. M. (2009). The yeast lysosome-like vacuole: Endpoint and crossroads. Biochimica et Biophysica Acta (BBA)-Molecular Cell Research, *1793*(4), 650-663.

[5] Graham, Laurie A., and Tom H. Stevens. "Assembly of the yeast vacuolar proton-translocating ATPase." Journal of bioenergetics and biomembranes 31.1 (1999): 39-47.

[6] Ohira, Masashi, et al. "The E and G subunits of the yeast V-ATPase interact tightly and are both present at more than one copy per V1 complex." Journal of Biological Chemistry 281.32 (2006): 22752-22760.

[7] Mellman, I., Fuchs, R., & Helenius, A. (1986). Acidification of the endocytic and exocytic pathways. Annual review of biochemistry, *55*(1), 663-700.

[8] Parra, K. J., and Kane, P. M. (1998). Reversible association between the V1 and V0 domains of yeast vacuolar H+-ATPase is an unconventional glucose-induced effect. Mol. Cell. Biol. 18, 7064-7074. doi: 10.1128/MCB.18.12.7064

[9] Forgac, M. (2007). Vacuolar ATPases: Rotary proton pumps in physiology and pathophysiology. Nat. Rev. Mol. Cell Biol. 8, 917-929. doi: 10.1038/nrm2272

[10] Davies, S. A., Goodwin, S. F., Kelly, D. C., Wang, Z., Sözen, M. A., Kaiser, K., & Dow, J. A. (1996). Analysis and inactivation of vha55, the gene encoding the vacuolar ATPase B-subunit in Drosophila melanogaster reveals a larval lethal phenotype. Journal of Biological Chemistry, *271*(48), 30677-30684.

[11] Sun-Wada, G. H., Murata, Y., Yamamoto, A., Kanazawa, H., Wada, Y., & Futai, M. (2000). Acidic endomembrane organelles are required for mouse postimplantation development. Developmental biology, *228*(2), 315-325.

[12] Kane, P. M. (2006). The where, when, and how of organelle acidification by the yeast vacuolar H+-ATPase. Microbiology and Molecular Biology Reviews, *70*(1), 177-191.

[13] Kane, Patricia M. "Close-up and genomic views of the yeast vacuolar H+-ATPase." Journal of bioenergetics and biomembranes 37.6 (2005): 399-403.

[14] Clemens, Stephan, et al. "Tolerance to toxic metals by a gene family of phytochelatin synthases from plants and yeast." The EMBO Journal 18.12 (1999): 3325-3333.

[15] Förster, Carola, and Patricia M. Kane. "Cytosolic Ca2+ homeostasis is a constitutive function of the V-ATPase in Saccharomyces cerevisiae." Journal of Biological Chemistry 275.49 (2000): 38245-38253.

[16] Ohya, Y., et al. "Calcium-sensitive cls mutants of Saccharomyces cerevisiae showing a Petphenotype are ascribable to defects of vacuolar membrane H (+)-ATPase activity." Journal of Biological Chemistry 266.21 (1991): 13971-13977.

[17] Bryant NJ, Piper RC, Gerrard SR, Stevens TH. Traffic into the prevacuolar/endosomal compartment of Saccharomyces cerevisiae: a VPS45-dependent intracellular route and a VPS45-independent, endocytic route. Eur J Cell Biol. 1998 May;76(1):43-52. doi: 10.1016/S0171-9335(98)80016-2. PMID: 9650782.

[18] Sousa CA, Perez RR, Soares EV. Saccharomyces cerevisiae mutants affected in vacuole assembly or vacuolar H+-ATPase are hypersensitive to lead (Pb) toxicity. Curr Microbiol. 2014 Jan;68(1):113-9. doi: 10.1007/s00284-013-0438-y. Epub 2013 Sep 8. PMID: 24013611.

[19] Yamashiro, C. T., et al. "Role of vacuolar acidification in protein sorting and zymogen activation: A genetic analysis of the yeast vacuolar proton-translocating ATPase." Molecular and Cellular Biology 10.7 (1990): 3737-3749.

[20] Nakamura, Norihiro, et al. "Acidification of vacuoles is required for autophagic degradation in the yeast, Saccharomyces cerevisiae." Journal of biochemistry 121.2 (1997): 338-344.

[21] Scott, Sidney V., et al. "Cytoplasm-to-vacuole targeting and autophagy employ the same machinery to deliver proteins to the yeast vacuole." Proceedings of the national academy of sciences 93.22 (1996): 12304-12308.

[22] Yorimitsu, T., and Daniel J. Klionsky. "Autophagy: Molecular machinery for self-eating." Cell Death & Differentiation 12 (2005): 1542-1552.

[23] Graham, Laurie A., Andrew R. Flannery, and Tom H. Stevens. "Structure and assembly of the yeast V-ATPase." Journal of bioenergetics and biomembranes 35.4 (2003): 301-312.

[24] Kawasaki-Nishi, S., Nishi, T., & Forgac, M. (2001). Yeast V-ATPase complexes containing different isoforms of the 100-kDa a-subunit differ in coupling efficiency and in VivoDissociation. Journal of Biological Chemistry, *276*(21), 17941-17948.

[25] Parra, K. J., Chan, C. Y., and Chen, J. (2014). *Saccharomyces cerevisiae* vacuolar H+-ATPase regulation by disassembly and reassembly: One structure and multiple signals. Eukaryot. Cell 13, 706-714. doi: 10.1128/EC.00050-14

[26] Shao, Elim, et al. "Mutational analysis of the non-homologous region of subunit a of the yeast VATPase." Journal of Biological Chemistry 278.15 (2003): 12985-12991.

[27] Wagner, Carsten A., et al. "Renal vacuolar H+-ATPase." Physiological Reviews 84.4 (2004): 1263-1314.

[28] Curtis, Kelly Keenan, et al. "Mutational analysis of the subunit C (Vma5p) of the yeast vacuolar H+-ATPase." Journal of Biological Chemistry 277.11 (2002): 8979-8988.

[29] Drory, Omri, Felix Frolow, and Nathan Nelson. "Crystal structure of yeast V-ATPase subunit C reveals its stator function." EMBO reports 5.12 (2004): 1148-1152.

[30] Xu, Ting, and Michael Forgac. "Subunit D (Vma8p) of the yeast vacuolar H+-ATPase plays a role in coupling of proton transport and ATP hydrolysis." Journal of Biological Chemistry 275.29 (2000): 22075-22081.

[31] Grüber, Gerhard, et al. "Expression, purification, and characterization of subunit E, an essential subunit of the vacuolar ATPase." Biochemical and biophysical research communications 298.3 (2002): 383-391.

[32] Jefferies, Kevin C., and Michael Forgac. "Subunit H of the vacuolar (H+) ATPase inhibits ATP hydrolysis by the free V1 domain by interaction with the rotary subunit F." Journal of Biological Chemistry 283.8 (2008): 4512-4519

[33] Lu, Ming, et al. "The amino-terminal domain of the E subunit of vacuolar H+-ATPase (V-ATPase) interacts with the H subunit and is required for V-ATPase function." Journal of Biological Chemistry 277.41 (2002): 38409-38415.

[34] Nelson, Hannah, Sreekala Mandiyan, and Nathan Nelson. "The Saccharomyces cerevisiae VMA7 gene encodes a 14-kDa subunit of the vacuolar H (+)-ATPase catalytic sector." Journal of Biological Chemistry 269.39 (1994): 24150-24155.

[35] Charsky, Colleen MH, Nicole J. Schumann, and Patricia M. Kane. "Mutational analysis of subunit G (Vma10p) of the yeast vacuolar H+-ATPase." Journal of Biological Chemistry 275.47 (2000): 37232-37239.

[36] Leng, Xing-Hong, et al. "Site-directed mutagenesis of the 100-kDa subunit (Vph1p) of the yeast vacuolar (H+)-ATPase." Journal of Biological Chemistry 271.37 (1996): 22487-22493.

[37] Owegi, Margaret A., et al. "Identification of a domain in the Vo subunit d that is critical for coupling of the yeast vacuolar proton-translocating ATPase." Journal of Biological Chemistry 281.40 (2006): 30001-30014.

[38] Umemoto, N., et al. "Roles of the VMA3 gene product, subunit c of the vacuolar membrane H (+)- ATPase on vacuolar acidification and protein transport. A study with VMA3-disrupted mutants of Saccharomyces cerevisiae." Journal of Biological Chemistry 265.30 (1990): 18447-18453.

[39] Bauerle, C., et al. "The Saccharomyces cerevisiae VMA6 gene encodes the 36-kDa subunit of the vacuolar H (+)-ATPase membrane sector." Journal of Biological Chemistry 268.17 (1993): 12749-12757.

[40] Bueler, S. A., & Rubinstein, J. L. (2015). Vma9p need not be associated with the yeast V-ATPase for fully-coupled proton pumping activity in vitro. Biochemistry, *54*(3), 853-858.

[41] Graham, LAURIE A., B. Powell, and T. H. Stevens. "Composition and assembly of the yeast vacuolar H (+)-ATPase complex." Journal of Experimental Biology 203.1 (2000): 61-70.

[42] Anraku, Yasuhiro, et al. "Genetic and cell biological aspects of the yeast vacuolar H+-ATPase." Journal of bioenergetics and biomembranes 24.4 (1992): 395-405.

[43] Kane, P. M. (1995). Disassembly and reassembly of the yeast vacuolar H(+)-ATPase in vivo. J. Biol. Chem. 270, 17025-17032.

[44] Tabke, K., Albertmelcher, A., Vitavska, O., Huss, M., Schmitz, H. P., and Wieczorek, H. (2014). Reversible disassembly of the yeast V-ATPase revisited under in vivo conditions. Biochem. J. 462, 185-197. doi: 10.1042/BJ20131293

[45] Cipriano, D. J., Wang, Y., Bond, S., Hinton, A., Jefferies, K. C., and Forgac, M. (2008). Structure and regulation of the vacuolar ATPases. Biochim. Biophys. Acta 1777, 599-604. doi: 10.1016/j.bbabio.2008.03.013

[46] Diab, H., Ohira, M., Liu, M., and Kane, P. M. (2009). Subunit interactions and requirements for inhibition of the yeast V1-ATPase. J. Biol. Chem. 284, 13316-13325. doi: 10.1074/jbc.M900475200.

[47] Parra, K. J., Keenan, K. L., and Kane, P. M. (2000). The H subunit

(Vma13p) of the yeast V-ATPase inhibits the ATPase activity of cytosolic V1 complexes. J. Biol. Chem. 275, 21761-21767. doi: 10.1074/jbc.M002305200

[48] Sharma, S., Oot, R. A., and Wilkens, S. (2018). MgATP hydrolysis destabilizes the interaction between subunit H and yeast V1-ATPase, highlighting H's role in V-ATPase regulation by reversible disassembly. J. Biol. Chem. 293, 10718-10730. doi: 10.1074/jbc.RA118.002951

[49] Couoh-Cardel, S., and Wilkens, S. (2015). Affinity purification and structural features of the yeast vacuolar ATPase Vo membrane sector. J. Biol. Chem. 290, 27959-27971. doi: 10.1074/jbc.M115.662494

[50] Smardon, A. M., Nasab, N. D., Tarsio, M., Diakov, T. T., and Kane, P. M. (2015). Molecular interactions and cellular itinerary of the yeast RAVE (regulator of the H+-ATPase of vacuolar and endosomal membranes) complex. J. Biol. Chem. 290, 27511-27523. doi: 10.1074/jbc.M115.667634

[51] Seol, J. H., Shevchenko, A., and Deshaies, R. J. (2001). Skp1 forms multiple protein complexes, including RAVE, a regulator of V-ATPase assembly. Nat. Cell Biol. 3, 384-391. doi: 10.1038/35070067

[52] Smardon, A. M., Tarsio, M., and Kane, P. M. (2002). The RAVE complex is essential for stable assembly of the yeast V-ATPase. J. Biol. Chem. 277, 13831-13839. doi: 10.1074/jbc.M200682200

[53] Smardon, A. M., and Kane, P. M. (2007). RAVE is essential for the efficient assembly of the C subunit with the vacuolar H(+)-ATPase. J. Biol. Chem. 282, 26185-26194. doi: 10.1074/jbc.M703627200

[54] Colacurcio, D. J., & Nixon, R. A. (2016). Disorders of lysosomal acidification—The emerging role of v-ATPase in aging and neurodegenerative disease. Ageing research reviews, *32*, 75-88.

[55] Nixon, R. A., & Yang, D. S. (2011). Autophagy failure in Alzheimer's disease—Locating the primary defect. Neurobiology of disease, *43*(1), 38-45.

[56] Zare-Shahabadi, A., Masliah, E., Johnson, G. V., & Rezaei, N. (2015). Autophagy in Alzheimer's disease. Reviews in the Neurosciences, *26*(4), 385-395.

[57] Dehay, B., Martinez-Vicente, M., Caldwell, G. A., Caldwell, K. A., Yue, Z., Cookson, M. R., ... & Bezard, E. (2013). Lysosomal impairment in Parkinson's disease. Movement Disorders, 28(6), 725-732.

[58] Boland, B., & Platt, F. M. (2015). Bridging the age spectrum of neurodegenerative storage diseases. Best Practice & Research Clinical Endocrinology & Metabolism, *29*(2), 127-143.

[59] Hughes, A. L., & Gottschling, D. E. (2012). An early age increase in vacuolar pH limits mitochondrial function and lifespan in yeast. Nature, *492*(7428), 261-265.

[60] Ruckenstuhl, C., Netzberger, C., Entfellner, I., Carmona-Gutierrez, D., Kickenweiz, T., Stekovic, S., ... & Madeo, F. (2014). Lifespan extension by methionine restriction requires autophagy-dependent vacuolar acidification. PLoS Genet, *10*(5), e1004347.

[61] Ballabio, A., & Gieselmann, V. (2009). Lysosomal disorders: From storage to cellular damage. Biochimica et Biophysica Acta (BBA)-Molecular Cell Research, *1793*(4), 684-696.

[62] Coutinho, M. F., & Alves, S. (2016). From rare to common and back again:

60 years of lysosomal dysfunction. Molecular genetics and metabolism, *117*(2), 53-65.

[63] Futerman, A. H., & Van Meer, G. (2004). The cell biology of lysosomal storage disorders. Nature reviews Molecular cell biology, *5*(7), 554-565.

[64] Hinton, A., Bond, S., & Forgac, M. (2009). V-ATPase functions in normal and disease processes. Pflügers archiv-European journal of physiology, *457*(3), 589-598.

[65] Hinton, A., Sennoune, S. R., Bond, S., Fang, M., Reuveni, M., Sahagian, G. G., ... & Forgac, M. (2009). Function of a subunit isoforms of the V-ATPase in pH homeostasis and in vitro invasion of MDA-MB231 human breast cancer cells. Journal of Biological Chemistry, *284*(24), 16400-16408.

[66] Carman, G. M., and S. A. Henry. "Phospholipid biosynthesis in yeast." Annual review of biochemistry 58.1 (1989): 635-667.

[67] Henry SA, Kohlwein SD, Carman GM (2012). Metabolism and regulation of Glycerolipids in the yeast Saccharomyces cerevisiae. Genetics, 190(2): 317-349.

[68] Van Meer, Gerrit, Dennis R. Voelker, and Gerald W. Feigenson. "Membrane lipids: Where they are and how they behave." Nature reviews molecular cell biology 9.2 (2008): 112-124.

[69] Boumann, H. A., Gubbens, J., Koorengevel, M. C., Oh, C. S., Martin, C. E., Heck, A. J., Patton-Vogt, J., Henry, S.A., de Kruijff, B., & de Kroon, A. I. (2006). Depletion of phosphatidylcholine in yeast induces shortening and increased saturation of the lipid acyl chains: Evidence for regulation of intrinsic membrane curvature in a eukaryote. Molecular biology of the cell, 17(2), 1006-1017.

[70] Dowd, S. R., Bier, M. E., & Patton-Vogt, J. L. (2001). Turnover of phosphatidylcholine in Saccharomyces cerevisiae THE ROLE OF THE CDP-CHOLINE PATHWAY. Journal of Biological Chemistry, 276(6), 3756-3763.

[71] Gaspar, M. L., Aregullin, M. A., Jesch, S. A., & Henry, S. A. (2006). Inositol induces a profound alteration in the pattern and rate of synthesis and turnover of membrane lipids in Saccharomyces cerevisiae. Journal of Biological Chemistry, 281(32), 22773-22785.

[72] Kelley, M. J., Bailis, A. M., Henry, S. A., & Carman, G. M. (1988). Regulation of phospholipid biosynthesis in Saccharomyces cerevisiae by inositol. Inositol is an inhibitor of phosphatidylserine synthase activity. Journal of Biological Chemistry, 263(34), 18078-18085.

[73] Bachhawat N, Ouyang Q, Henry SA (1995). Functional characterization of an Inositolsensitive upstream activation sequence in yeast. J Biol Chem, 270 (42): 25087-25095.

[74] Carman GM and Henry SA (1999). Phospholipid biosynthesis in the yeast Saccharomyces cerevisiae and interrelationship with other metabolic processes. Prog Lipid Res., 38(5-6):361-399.

[75] Carman GM and Han GS (2011). Regulation of phospholipid synthesis in the yeast Saccharomyces cerevisiae. Annu Rev Biochem, 80:858-883.

[76] Han GS, Siniossoglou S, Carman GM (2007). The cellular functions of the yeast lipin homolog PAH1p are dependent on its phosphatidate phosphatase activity. J Biol Chem, 282:37026-37035.

[77] Pascual, F., and Carman, G. M. (2012). Phosphatidate phosphatase, a key regulator of lipid homeostasis.

Biochimica et Biophysica Acta (BBA)-molecular and cell biology of lipids.

[78] Carman GM and Han GS (2006). Roles of phosphatidate phosphatase enzymes in lipid metabolism. Trends Biochem Sci., 31(12):694-699.

[79] Toke, D. A., Bennett, W. L., Oshiro, J., Wu, W. I., Voelker, D. R., & Carman, G. M. (1998). Isolation and characterization of the Saccharomyces cerevisiae LPP1 gene encoding a Mg2+-independent phosphatidate phosphatase. Journal of Biological Chemistry, 273(23), 14331-14338.

[80] Han GS, Wu WI, Carman GM (2006). The Saccharomyces cerevisiae Lipin homolog is a Mg2+-dependent phosphatidate phosphatase enzyme. J Biol Chem, 281:9210-9218.

[81] Sorger, D., & Daum, G. (2003). Triacylglycerol biosynthesis in yeast. Applied microbiology and biotechnology, 61(4), 289-299.

[82] Karanasios E, Han GS, Carman GM, Siniossoglou S (2010). A phosphorylation-regulated amphipathic helix controls the membrane translocation and function of the yeast phosphatidate phosphatase. PNAS, 107: 17539-17544.

[83] Santos-Rosa H, Leung J, Grimsey N, Peak-Chew S, Siniossoglou S (2005). The yeast lipin Smp2 couples phospholipid biosynthesis to nuclear membrane growth. EMBO J, 24:1931-1941.

[84] Wimalarathna R, Tsai CH, Shen CH (2011). Transcriptional control of genes involved in yeast phospholipid biosynthesis. JMicrobiol, 49(2):265-273.

[85] Loewen, C. J. R., Gaspar, M. L., Jesch, S. A., Delon, C., Ktistakis, N. T., Henry, S. A., & Levine, T. P. (2004). Phospholipid metabolism regulated by a transcription factor sensing phosphatidic acid. Science, 304(5677), 1644-1647.

[86] Siniossoglou S (2009). Lipins, lipids and nuclear envelope structure. Traffic, 10(9):1181-1187

[87] Soto-Cardalda A, Fakas S, Pascual F, Choi HS, Carman GM (2012). Phosphatidate phosphatase plays role in zinc-mediated regulation of phospholipid synthesis in yeast. J biol Chem, 6;287(2):968-77.

[88] Barbosa, A. D., Sembongi, H., Su, W. M., Abreu, S., Reggiori, F., Carman, G. M., & Siniossoglou, S. (2015). Lipid Partitioning at the Nuclear Envelope Controls Membrane biogenesis. Molecular biology of the cell, 26(20), 3641-3657.

[89] Fakas S, Qiu Y, Dixon JL, Han GS, Ruggles KV, Garbarino J, Sturley SL, Carman GM (2011). Phosphatidate phosphatase activity plays key role in protection against fatty acidinduced toxicity in yeast. J Biol Chem, 286(33) 29074-29085.

[90] Adeyo, O., Horn, P. J., Lee, S., Binns, D. D., Chandrahas, A., Chapman, K. D., & Goodman, J. M. (2011). The yeast lipin orthologue Pah1p is important for biogenesis of lipid droplets.The Journal of cell biology, 192(6), 1043-1055.

[91] Hsieh, L. S., Su, W. M., Han, G. S., & Carman, G. M. (2015). Phosphorylation regulates the ubiquitin-independent degradation of yeast Pah1 phosphatidate phosphatase by the 20S proteasome. Journal of Biological Chemistry, 290(18), 11467-11478.

[92] Sasser, T., Qiu, Q. S., Karunakaran, S., Padolina, M., Reyes, A., Flood, B. and Fratti, R. A. (2012). Yeast lipin 1 orthologue pah1p regulates vacuole homeostasis and membrane fusion. Journal of Biological Chemistry, 287(3), 2221-2236.

[93] Sherr, Goldie Libby, et al. "Pah1p negatively regulates the expression of V-ATPase genes as well as vacuolar acidification." Biochemical and biophysical research communications 491.3 (2017): 693-700.

[94] Conradt, Barbara, et al. "In vitro reactions of vacuole inheritance in Saccharomyces cerevisiae." The Journal of cell biology 119.6 (1992): 1469-1479.

[95] Wiemken, A., Ph Matile, and H. Moor. (1970). "Vacuolar dynamics in synchronously budding yeast." Archiv für Mikrobiologie 70.2: 89-103.

[96] Desfougères, Y., Vavassori, S., Rompf, M., Gerasimaite, R., & Mayer, A. (2016). Organelle acidification negatively regulates vacuole membrane fusion in vivo. Scientific reports, *6*(1), 1-16.

[97] Raymond, C. K., Howald-Stevenson, I., Vater, C. A., & Stevens, T. H. (1992). Morphological classification of the yeast vacuolar protein sorting mutants: Evidence for a prevacuolar compartment in class E vps mutants. Molecular biology of the cell, *3*(12), 1389-1402.

[98] Weisman, L. S., Bacallao, R., & Wickner, W. (1987). Multiple methods of visualizing the yeast vacuole permit evaluation of its morphology and inheritance during the cell cycle. Journal of Cell Biology, *105*(4), 1539-1547.

[99] Sautin, Y. Y., Lu, M., Gaugler, A., and Gluck, S. L. (2005). Phosphatidylinositol 3-kinase-mediated effects of glucose on vacuolar H+-ATPase assembly, translocation, and acidification of intracellular compartments in renal epithelial cells. Mol. Cell. Biol. 25, 575-589. doi: 10.1128/MCB.25.2.575-589.2005

[100] Marjuki, H., Gornitzky, A., Marathe, B. M., Ilyushina, N. A., Aldridge, J. R., Desai, G., et al. (2011). Influenza a virus-induced early activation of ERK and PI3K mediates V-ATPase-dependent intracellular pH change required for fusion. Cell. Microbiol. 13, 587-601. doi: 10.1111/j.1462-5822.2010.01556.x

[101] Chan, C. Y., and Parra, K. J. (2014). Yeast phosphofructokinase-1 subunit Pfk2p is necessary for pH homeostasis and glucose-dependent vacuolar ATPase reassembly. J. Biol. Chem. 289, 19448-19457. doi: 10.1074/jbc.M114.569855

[102] Chan, C. Y., and Parra, K. J. (2016). Regulation of vacuolar H+-ATPase (V-ATPase) reassembly by glycolysis flow in 6-phosphofructo-1-kinase (PFK-1)-deficient yeast cells. J. Biol. Chem. 291, 15820-15829. doi: 10.1074/jbc.M116.717488

[103] Bond, S., and Forgac, M. (2008). The Ras/cAMP/protein kinase a pathway regulates glucose-dependent assembly of the vacuolar (H+)-ATPase in yeast. J. Biol. Chem. 283, 36513-36521. doi: 10.1074/jbc.M805232200

[104] Dechant, R., Binda, M., Lee, S. S., Pelet, S., and Peter, M. (2010). Cytosolic pH is a second messenger for glucose and regulates the PKA pathway through V-ATPase. EMBO J. 29, 2515-2526. doi: 10.1038/emboj.2010.138

[105] Purwin, C., Nicolay, K., Scheffers, W. A., and Holzer, H. (1986). Mechanism of control of adenylate cyclase activity in yeast by fermentable sugars and carbonyl cyanide m-chlorophenylhydrazone. J. Biol. Chem. 261, 8744-8749.

[106] Stransky, L., Cotter, K., & Forgac, M. (2016). The function of V-ATPases in cancer. Physiological reviews, *96*(3), 1071-1091.

[107] Anastas, J. N., & Moon, R. T. (2013). WNT signalling pathways as therapeutic targets in cancer. Nature Reviews Cancer, *13*(1), 11-26.

[108] D'Angelo, R. C., Ouzounova, M., Davis, A., Choi, D., Tchuenkam, S. M., Kim, G., ... & Korkaya, H. (2015). Notch reporter activity in breast cancer cell lines identifies a subset of cells with stem cell activity. Molecular cancer therapeutics, *14*(3), 779-787.

[109] Laplante, M., & Sabatini, D. M. (2012). mTOR signaling in growth control and disease. Cell, *149*(2), 274-293.

[110] Xu, Y., Parmar, A., Roux, E., Balbis, A., Dumas, V., Chevalier, S., & Posner, B. I. (2012). Epidermal growth factor-induced vacuolar (H+)-atpase assembly: A role in signaling via mTORC1 activation. Journal of biological chemistry, *287*(31), 26409-26422.

[111] Xu, J., Xie, R., Liu, X., Wen, G., Jin, H., Yu, Z., ... & Tuo, B. (2012). Expression and functional role of vacuolar H+-ATPase in human hepatocellular carcinoma. Carcinogenesis, *33*(12), 2432-2440.

[112] Zhang, C. S., Jiang, B., Li, M., Zhu, M., Peng, Y., Zhang, Y. L., ... & Lin, S. C. (2014). The lysosomal v-ATPase-Ragulator complex is a common activator for AMPK and mTORC1, acting as a switch between catabolism and anabolism. Cell metabolism, *20*(3), 526-540.

[113] Zoncu, R., Bar-Peled, L., Efeyan, A., Wang, S., Sancak, Y., & Sabatini, D. M. (2011). mTORC1 senses lysosomal amino acids through an inside-out mechanism that requires the vacuolar H+-ATPase. Science, *334*(6056), 678-683.

[114] McHenry, P., Wang, W. L. W., Devitt, E., Kluesner, N., Davisson, V. J., McKee, E., ... & Tenniswood, M. (2010). Iejimalides a and B inhibit lysosomal vacuolar H+-ATPase (V-ATPase) activity and induce S-phase arrest and apoptosis in MCF-7 cells. Journal of cellular biochemistry, *109*(4), 634-642.

[115] Morimura, T., Fujita, K., Akita, M., Nagashima, M., & Satomi, A. (2008). The proton pump inhibitor inhibits cell growth and induces apoptosis in human hepatoblastoma. Pediatric surgery international, *24*(10), 1087-1094.

[116] Nishihara, T., Akifusa, S., Koseki, T., Kato, S., Muro, M., & Hanada, N. (1995). Specific inhibitors of vacuolar type H+-ATPases induce apoptotic cell death. Biochemical and biophysical research communications, *212*(1), 255-262.

[117] Ohta, T., Arakawa, H., Futagami, F., Fushida, S., Kitagawa, H., Kayahara, M., ... & Ohkuma, S. (1998). Bafilomycin A1 induces apoptosis in the human pancreatic cancer cell line Capan-1. The Journal of Pathology: A Journal of the Pathological Society of Great Britain and Ireland, *185*(3), 324-330.

[118] Schempp, C. M., von Schwarzenberg, K., Schreiner, L., Kubisch, R., Müller, R., Wagner, E., & Vollmar, A. M. (2014). V-ATPase inhibition regulates anoikis resistance and metastasis of cancer cells. Molecular cancer therapeutics, *13*(4), 926-937.

[119] Von Schwarzenberg, K., Wiedmann, R. M., Oak, P., Schulz, S., Zischka, H., Wanner, G., ... & Vollmar, A. M. (2013). Mode of cell death induction by pharmacological vacuolar H+-ATPase (V-ATPase) inhibition. Journal of Biological Chemistry, *288*(2), 1385-1396.

[120] Damaghi, M., Wojtkowiak, J. W., & Gillies, R. J. (2013). pH sensing and regulation in cancer. Frontiers in physiology, *4*, 370.

[121] Neri, D., & Supuran, C. T. (2011). Interfering with pH regulation in tumours as a therapeutic strategy. Nature reviews Drug discovery, *10*(10), 767-777.

[122] Bernard, D., Gebbia, M., Prabha, S., Gronda, M., MacLean, N., Wang, X., ... & Schimmer, A. D. (2015). Select microtubule inhibitors increase lysosome acidity and promote lysosomal disruption in acute myeloid leukemia (AML) cells. Apoptosis, *20*(7), 948-959.

[123] Capecci, J., & Forgac, M. (2013). The function of vacuolar ATPase (V-ATPase) a subunit isoforms in invasiveness of MCF10a and MCF10CA1a human breast cancer cells. Journal of Biological Chemistry, *288*(45), 32731-32741.

[124] Sennoune, S. R., Bakunts, K., Martínez, G. M., Chua-Tuan, J. L., Kebir, Y., Attaya, M. N., & Martínez-Zaguilán, R. (2004). Vacuolar H+-ATPase in human breast cancer cells with distinct metastatic potential: Distribution and functional activity. American journal of physiology-cell physiology, *286*(6), C1443-C1452.

[125] Chung, C., Mader, C. C., Schmitz, J. C., Atladottir, J., Fitchev, P., Cornwell, M. L., ... & Gorelick, F. (2011). The vacuolar-ATPase modulates matrix metalloproteinase isoforms in human pancreatic cancer. Laboratory investigation, *91*(5), 732-743.

[126] Huang, L., Lu, Q., Han, Y., Li, Z., Zhang, Z., & Li, X. (2012). ABCG2/V-ATPase was associated with the drug resistance and tumor metastasis of esophageal squamous cancer cells. Diagnostic Pathology, *7*(1), 1-7.

[127] Kulshrestha, A., Katara, G. K., Ibrahim, S., Pamarthy, S., Jaiswal, M. K., Sachs, A. G., & Beaman, K. D. (2015). Vacuolar ATPase 'a2'isoform exhibits distinct cell surface accumulation and modulates matrix metalloproteinase activity in ovarian cancer. Oncotarget, *6*(6), 3797.

[128] Lu, Q., Lu, S., Huang, L., Wang, T., Wan, Y., Zhou, C. X., ... & Li, X. (2013). The expression of V-ATPase is associated with drug resistance and pathology of non-small-cell lung cancer. Diagnostic pathology, *8*(1), 1-7.

[129] Michel, V., Licon-Munoz, Y., Trujillo, K., Bisoffi, M., & Parra, K. J. (2013). Inhibitors of vacuolar ATPase proton pumps inhibit human prostate cancer cell invasion and prostate-specific antigen expression and secretion. International journal of cancer, *132*(2), E1-E10.

[130] De Milito, A., Canese, R., Marino, M. L., Borghi, M., Iero, M., Villa, A., ... & Fais, S. (2010). pH-dependent antitumor activity of proton pump inhibitors against human melanoma is mediated by inhibition of tumor acidity. *International journal of cancer*, *127*(1), 207-219.

[131] Lee, G. H., Kim, D. S., Kim, H. T., Lee, J. W., Chung, C. H., Ahn, T., ... & Kim, H. R. (2011). Enhanced lysosomal activity is involved in Bax inhibitor-1-induced regulation of the endoplasmic reticulum (ER) stress response and cell death against ER stress: Involvement of vacuolar H+-ATPase (V-ATPase). Journal of Biological Chemistry, *286*(28), 24743-24753.

[132] Villa-García, M. J., Choi, M. S., Hinz, F. I., Gaspar, M. L., Jesch, S. A., & Henry, S. A. (2011). Genome-wide screen for inositol auxotrophy in Saccharomyces cerevisiae implicates lipid metabolism in stress response signaling. Molecular genetics and genomics, *285*(2), 125-149.

[133] Young, B. P., Shin, J. J., Orij, R., Chao, J. T., Li, S. C., Guan, X. L., ... & Loewen, C. J. (2010). Phosphatidic acid is a pH biosensor that links membrane biogenesis to metabolism. Science, *329*(5995), 1085-1088.

[134] Konarzewska, P., Sherr, G. L., Ahmed, S., Ursomanno, B., & Shen, C. H. (2017). Vma3p protects cells from programmed cell death through the regulation of Hxk2p expression.

Biochemical and biophysical research communications, *493*(1), 233-239.

[135] Amigoni, L., Martegani, E., & Colombo, S. (2013). Lack of HXK2 induces localization of active Ras in mitochondria and triggers apoptosis in the yeast Saccharomyces cerevisiae. Oxidative medicine and cellular longevity, *2013*.

[136] Palomino, A., Herrero, P., & Moreno, F. (2005). Rgt1, a glucose sensing transcription factor, is required for transcriptional repression of the HXK2 gene in Saccharomyces cerevisiae. Biochemical journal, *388*(2), 697-703.

[137] Li, S. C., Diakov, T. T., Xu, T., Tarsio, M., Zhu, W., Couoh-Cardel, S., ... & Kane, P. M. (2014). The signaling lipid PI (3, 5) P2 stabilizes V1–Vo sector interactions and activates the V-ATPase. Molecular biology of the cell, *25*(8), 1251-1262.

[138] Hu, Y., Lu, W., Chen, G., Wang, P., Chen, Z., Zhou, Y., ... & Huang, P. (2012). K-ras G12V transformation leads to mitochondrial dysfunction and a metabolic switch from oxidative phosphorylation to glycolysis. Cell research, *22*(2), 399-412.